# EQUIPMENT MANAGEMENT
## WORKBOOK

## Key to Equipment Reliability and Productivity in Mining

Paul D. Tomlingson

**Published by**

Society for
Mining, Metallurgy
& Exploration

**Your most
precious resource.**

Society for Mining, Metallurgy, and Exploration, Inc. (SME)
8307 Shaffer Parkway
Littleton, Colorado, USA 80127
(303) 948-4200 / (800) 763-3132
www.smenet.org

SME advances the worldwide mining and minerals community through information exchange and professional development. With members in more than 71 countries, SME is the world's largest association of mining and minerals professionals.



ISBN-13: 978-0-87335-335-9

**Library of Congress Cataloging-in-Publication Data**
Tomlingson, Paul D.
Equipment mangement workbook : key to equipment reliability and productivity in mining / by Paul D. Tomlingson.
     p. cm.
ISBN 978-0-87335-335-9
1. Mining machinery--Maintenance and repair--Handbooks, manuals, etc. 2. Mine management--Handbooks, manuals, etc. I. Title.
TN345.T63 2010
622.068'2--dc22

2010030492

# Contents

# Introduction

This workbook provides an opportunity for the reader of *Equipment Management: Key to Equipment Reliability and Productivity in Mining, 2nd ed.*, to review and confirm the lessons and recommendations gained from that text, which has the goal of offering a practical and effective strategy for ensuring the profitability of mining operations with quality maintenance management.

Collectively, the textbook and workbook provide a powerful and useful combination to assist mining organizations in the successful implementation of an equipment management strategy to better ensure realization of equipment reliability and work force productivity. The step-by-step presentation focuses on the most critical aspects of successful mining maintenance management. Each chapter in the textbook is based on real-world experiences and gives practical, realistic recommendations that can be put to immediate use. The reader is encouraged to read the introduction to the textbook and to examine the maintenance performance evaluation (MPE), which details the standards necessary to achieve "world-class" performance.

The *Equipment Management* textbook also provides a glossary defining all of the terms that are used in both the textbook and the workbook.

In addition to use by individual readers, the workbook can be utilized by mining organizations when conducting in-house training based on the textbook.

## USING THE WORKBOOK

In this text, questions are posed in several different ways: multiple choice, true/false, and yes/no. The reader is also asked to number listed items in logical order or fill in the blanks of a statement. **Where multiple-choice questions are given, more than one answer may be correct, so all applicable answers should be selected.**

Together with solutions to the questions, brief comments are offered with the answers where applicable, and pages in the textbook are cross-referenced. The workbook appendix provides a brief overview of the MPE.

# CHAPTER 1

# Understanding Equipment Management

1.  Equipment management focuses on the improvement of maintenance performance by implementing:

    a.  new technologies

    b.  responsive organizations

    c.  modern management skills

2.  The plant manager's production strategy includes:

    a.  department objectives

    b.  the company mission statement

    c.  policies

    d.  terminology

3.  Department objectives in the production strategy are amplified by policies. How do policies help implement the equipment management strategy?

    a.  policies guide development of procedures

    b.  procedures specify internal and interdepartmental actions

    c.  all of the above

4.  The maintenance program consists of nine elements given randomly in the following list. Number each element so that a logical sequence of program elements results.

    _____ plan work

    _____ measure work

    _____ assign work

    _____ schedule work

    _____ control work

    _____ identify work

    _____ request work

    _____ assess accomplishments

    _____ classify work

5.  Should the objectives assigned to each department contain elements prescribing interaction with other departments, as well as tasks performed by the department to which the objectives are assigned?

    ☐ Yes

    ☐ No

6.  What is characteristic of departmental objectives?

    a.  assigns tasks to departments

    b.  provides for mutual support among departments

    c.  ensures corresponding cooperative actions

    d.  can minimize surprises

7.  A policy for the conduct of preventive maintenance (PM) services might state:

    "The overall PM program will be assessed annually to ensure it covers all equipment requiring services and that the most appropriate types of services are applied at correct intervals. The performance of the PM program in reducing equipment failures and extending equipment life will be verified."

    Which of the following situations could this policy help to implement successfully?

    a.  maintenance has responsibility for establishing the program

    b.  maintenance would be guided on how to conduct the program

    c.  the policy would verify the plant manager's understanding of preventive maintenance

    d.  the plant manager would take steps to ensure PM effectiveness

# CHAPTER 2

# Applying the Principles of Equipment Management

1. What are some of the reasons why the basic principles of maintenance management have application in equipment management?

   a. equipment has become more complex and difficult to maintain

   b. greater technical skills are required to maintain equipment

   c. direct participation of all departments is necessary

   d. new strategies and revised programs are required

   e. higher quality information is necessary

   f. advanced condition monitoring is replacing time-based preventive maintenance

   g. all of the above

2. Why must the maintenance program be defined and made familiar to all plant personnel?

   a. the program may only prescribe internal maintenance actions

   b. the program may prescribe mutual interaction of all plant departments

   c. the program requires cooperation of all plant departments

3.  What are the principal reasons why maintenance should carefully identify and define the types of work they perform?

    a.  required work will be more expeditiously handled

    b.  work that maintenance cannot perform will not be requested

    c.  work force size and composition can be determined

    d.  sensible limitations can be placed on certain work

    e.  customer expectations can be defined

4.  Why should maintenance terminology be defined?

    a.  to avoid confusion among departments working with maintenance

    b.  to allow customers to accurately define their needs

    c.  to provide a common language throughout the whole plant

    d.  to allow a better understanding of maintenance communications

5.  Why should the maintenance organization not be determined until the maintenance program has been properly defined? Use the following words to fill in the answer blanks below: *defines, what, why.*

    The maintenance program _________ what maintenance must do, who will do _____, how they will do it, and _____.

6.  What are the expected results of effective labor control?

    a.  materials are installed more efficiently

    b.  quality work is done and repairs that have been made last longer

    c.  better labor use and reduced consumption of materials

    d.  maintenance costs are reduced

7.  Why should productivity be measured on a regular basis?

   a.  productivity is an indicator of maintenance effectiveness

   b.  improved productivity indicates better planning

   c.  high productivity is an element of effective work control

   d.  measuring productivity is a popular maintenance activity

8.  There would be little effective planning without a quality detection-oriented PM program.

   ☐ True

   ☐ False

9.  Why should there be criteria specifying which type of work should be planned and scheduled?

   a.  so that the most important jobs will be planned and scheduled

   b.  so that jobs not requiring planning will go directly to supervisors

   c.  supervision can focus on control of work

10.  What benefits are realized by applying standards to planned work?

   a.  standards provide details for repetitive jobs

   b.  planners can work more consistently and effectively

   c.  the work load can be determined

   d.  resource needs are refined with repetitive applications

11.  All major planned work should be jointly scheduled with operations.

   ☐ True

   ☐ False

12. What are the most essential aspects of a good maintenance priority-setting procedure?

    a. determining the relative priority of work

    b. identifying the time within which the work should be completed

    c. allowing operations to communicate work criticality

    d. enabling maintenance to indicate the importance of the work

    e. providing operations with expectation of work completion

    f. avoiding arguments

13. What are the essential functions of an information system?

    a. identifying the work

    b. controlling the work

    c. measuring the work

    d. converting field data into information

    e. making well-informed decisions

    f. all of the above

14. Maintenance depends on material control departments for its material needs but it does not control them. How can maintenance ensure the most effective support from material control departments?

    a. verify that the material control department's objectives include support of maintenance

    b. explain the maintenance program to the material control department

    c. include the material control department in scheduling sessions

    d. establish performance indices that depict performance

    e. all of the above

15. Nonmaintenance work such as equipment installation should be assessed to ensure it is necessary, feasible, and correctly funded before being assigned to maintenance. What are some of the consequences if this step is not taken?

    a. work could undermine the ability to carry out basic maintenance

    b. poor engineering could result in unsatisfactory work

    c. work could be incorrectly expensed rather than capitalized

    d. the maintenance budget could be exceeded

16. A maintenance evaluation should examine all internal and interdepartmental factors that affect maintenance performance.

    ☐ True

    ☐ False

# Developing the Equipment Management Program

1. Program definition begins with the plant manager, who:

    a. matches his production strategy with the company mission

    b. assigns department objectives and specifies how all departments work together

    c. provides policies to guide department interactions

    d. all of the above

2. What constitutes an equipment management program?

    a. the production strategy

    b. application of basic principles

    c. definition of the maintenance program

    d. all of the above

3. The maintenance program defines how work will be requested, identified, classified, planned, scheduled, assigned, controlled, measured, and the total effort evaluated. In the following list, match the departments (left column, by letter) with the elements of the program (right column) in which they are most often called to perform or participate. (NOTE: The departments may match more than one element.)

| Department | Element |
|---|---|
| a. operations | classifying ______ |
| b. maintenance | requesting ______ |
| c. purchasing | planning ______ |
| d. warehousing | measuring ______ |
| e. accounting | scheduling ______ |
| f. engineering | controlling ______ |

4. By defining the maintenance program management:

    a. existing program elements are confirmed

    b. new program elements can be developed more efficiently

    c. greater participation in change is encouraged

    d. commitments are made to support program objectives

    e. interactions among plant departments are clarified

    f. all of the above

5. Program education includes:

    a. an explanation of the program

    b. encouragement of questions and recommendations

    c. participation of maintenance, operations, and staff

    d. observation of discussions by managers, who reassert the need for cooperation

    e. all of the above

6. Why document the maintenance program using a schematic diagram?

    a.  it depicts interaction between participating departments

    b.  it shows "you" and "me" actions

    c.  it encourages participation

    d.  the accompanying legend aids understanding

    e.  all of the above

7. Initial program development should welcome the participation of as many personnel as possible. Why?

    a.  it encourages constructive suggestions

    b.  it creates a sense of ownership in the final program

    c.  it gains commitment to ensure program success

    d.  it prepares personnel to make minor program adjustments

    e.  all of the above

8. *Practical Exercise*

General
    On an 11" × 17" sheet of paper, use the following steps to draw a schematic diagram of the equipment or maintenance management program carried out in your plant.

1. Key Personnel and Activities

    First, list all of the key personnel or activities:
    a.  For example, key maintenance personnel to include superintendent, planner, supervisor, and craftsmen.

    b.  Principal operating personnel to include superintendent, supervisor, equipment operator, and scheduler, as appropriate.

    c.  Staff organizations, such as warehouse, purchasing, or accounting, for instance.

    d.  Plant shops that might support maintenance even if in the maintenance organization.

    e.  Support services such as transportation, cranes, rigging, and so forth.

    f.  Outside support services such as contractors or vendor shops performing component rebuild services.

    g.  Engineering to include plant engineering, process engineering, or mine engineering.

    h.  Key managers, such as the plant manager, general manager, or vice president of operations.

If you use a computer, or plan to include them in your future program, place a symbol for the computers near the personnel who might use them within the network. (Refer to symbols defined in the following section.)

2.  Technique

In pencil, place the key personnel in logical positions on your sheet of paper. By using pencil, your placements can be erased and moved to ensure clarity as the schematic is developed.

    Start by designing the PM program to include the following steps:

    a.  Determine PM services due next week.

    b.  Establish a plan for carrying out services.

    c.  Develop a schedule for services requiring equipment shutdown.

    d.  Provide supervisor with schedule.

    e.  Provide supervisor a list of services not requiring shutdown.

    f.  Show how supervisor makes assignments to craftsmen.

    g.  Show how craftsmen carry out PM services.

    h.  Show how craftsmen might confer with operators.

    i.  Show how equipment deficiencies are recorded and classified.

    j.  Show how operations is advised of new problems.

k.  Show how deficiencies are converted into work orders.

l.  Show how compliance with the PM schedule is measured.

m.  Describe the how success of preventive maintenance is determined.

As you design the program, use the following symbols:

- A computer

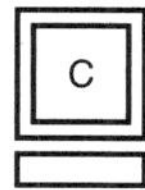

- A supervisor assigning work to a craftsman

  (Every action will subsequently be explained in a legend. Number the actions sequentially so that the reader of the diagram can follow the sequence [1, 2, 3, 4, etc.] and then read the text in the legend describing the action. Record the legend explanation number within a circle.)

- A craftsman conferring with an operator

- A craftsman reporting labor using an off-page connector (B)

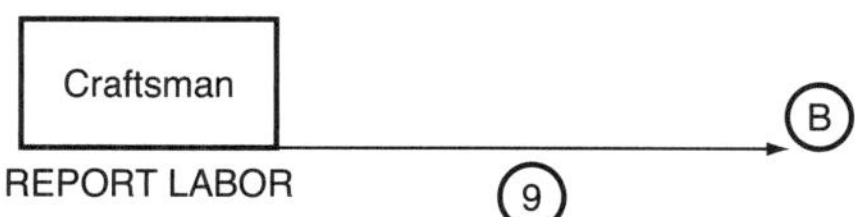

- Always try to draw lines horizontally or vertically. If one line must cross another, use a break. To illustrate:

Other parts of the program will subsequently be added. These might include planning steps, conducting periodic scheduling meetings, preparing for a shutdown, obtaining approval for new construction work, coordinating with contractors, and so forth. (Refer to pages 26–27 in the textbook.)

# CHAPTER 4

# Leadership in Equipment Management

Implementation of equipment management will influence actions and roles relative to conducting maintenance at every level of the corporate and plant organizations. Select the most likely actions or roles taken by the organization levels shown.

1. Corporate executives:

   a. examine potential for improvement offered by equipment management

   b. determine how equipment management can lower operating costs

   c. identify roles and responsibilities to support equipment management

   d. ensure plants have mastered maintenance fundamentals if effective maintenance

   e. verify effectiveness of information system

2. Plant managers:

   a. ensure that fundamentals of maintenance are being applied

   b. verify that the maintenance program is working effectively

   c. verify productivity of workers

   d. ensure maximum planned work

   e. verify steps to minimize downtime

   f. ensure responsible fiscal control

3. Maintenance managers:

   a. ensure mainenance supervisors use information effectively

   b. establish a quality maintenance engineering function

   c. ensure effective application of condition-monitoring techniques

   d. focus on productivity improvements

   e. educate the work force on expectations upon adoption of equipment management

4. Operations managers:

   a. prepare for implementation of business units

   b. become familiar with different maintenance concepts

   c. establish effective liaison with the maintenance manager

5. Staff departments:

   a. become familiar with maintenance program

   b. make adjustments in procedures to accommodate new responsibilities

   c. align service procedures with equipment management concepts

6. Maintenance superintendents:

    a. ensure that all plant personnel are familiar with program

    b. give emphasis to preventive maintenance

    c. verify utilization of effective technology

    d. organize to ensure effective control of labor

    e. utilize maintenance engineering to ensure program continuity

    f. coordinate regularly with production managers

7. Maintenance supervisors:

    a. ensure crews are familiar with program

    b. emphasize effective control of work

    c. focus on quality PM services

    d. ensure correct reporting of field data

    e. create a positive work environment

8. Maintenance engineers:

    a. verify procedures to gain reliability and maintainability

    b. review the conduct of PM services

    c. apply advanced technologies to PM program

    d. develop standards for periodic planned work

9. Maintenance planners:

    a. provide direct support in planning major jobs

    b. utilize criteria for planning work

    c. apply planning techniques

    d. utilize information for planning control

    e. work with the maintenance engineer

# CHAPTER 5

# Organization

1. Under what circumstances should a centrally controlled organization be utilized?

    a. when flexibility and responsiveness are not important

    b. to perform specialized functions

    c. to control uniquely qualified personnel

    d. all of the above

2. Which of the following actions are best centrally controlled?

    a. periodic technical predictive maintenance services

    b. lubrication

    c. instrumentation calibration

    d. emergency repairs

3. What are some of the characteristics of the craft organization?

    a. it assigns personnel of the same craft to a single supervisor

    b. it may require other crafts to complete complex multicraft jobs

    c. it is good at promoting craft skills

    d. it is highly responsive and flexible

4.  What are some of the limitations of the craft organization?

    a.  it must obtain outside craft skills for complex multicraft jobs

    b.  it could cause separation of mechanical versus electrical skills

    c.  it can create significant delays when the other crafts are needed

    d.  all of the above

5.  Match the types of organization in the left column (by letter) with the most correct description in the right column.

    a.  area organization        _____ requests come to one point

    b.  functional organization

    c.  centrally controlled        _____ poor multicraft coordination

    d.  craft organization

    _____ one supervisor responsible

    _____ performs specific tasks only

6.  What are some of the characteristics of the area organization?

    a.  it reduces problems of craft jurisdiction

    b.  repairs requiring several crafts can be dealt with more easily

    c.  greater flexibility and responsiveness are possible

    d.  productivity is likely to be higher than with craft organizations

7. Which of the following best describe a business unit?

     a. one person is responsible for both operations and maintenance

     b. it reduces "finger pointing" between operations and maintenance

     c. it provides opportunity for a maintenance–operations teamwork

     d. operators become more proficient in identifying problems

8. Which of the following should the maintenance planner avoid?

     a. getting parts for emergencies

     b. planning major jobs

     c. acting as relief supervisor

     d. planning construction work

     e. scheduling PM services not requiring shutdown

     f. controlling spare parts

     g. preparing daily work plans for supervisors

9. What are some of the most effective ways to reduce resistance to organizational change?

     a. educate personnel on the need for anticipated changes

     b. encourage participation in the change process

     c. allow time to absorb changes

     d. provide job security while changes are being made

     e. all of the above

10. Which conditions might inhibit team implementation?

    a. the maintenance program has not been clearly defined

    b. former supervisors have not been given new roles

    c. team members have little training in work control

    d. all of the above

11. A rotating team coordinator is a good interim step in converting from a traditional organization to a team because:

    a. it helps make the transition into a team organization

    b. there is opportunity to suggest, try, and discuss new work control techniques

    c. team members can rotate through coordinator positions

    d. it helps to standardize procedures

12. When supervisors are phased out, why must they be given new, positive tasks?

    a. the supervisor may resist the change

    b. the supervisor might feel redundant

    c. the supervisor could try to preserve the status quo

    d. the supervisor's actions could delay implementation

13. What decisions should team members make?

    a. when to take vacations

    b. how to perform work

    c. how to modify equipment after approval by engineering

    d. who gets overtime

    e. scheduling PM services after coordination with operations

    f. whether work will be done

14. Why should there be a well-defined program before team implementation is attempted?

     a. new team members need solid work control guidelines

     b. supervisors are no longer available for work control

     c. none of the above

# CHAPTER 6

# Work Load Versus Work Force

1. The work load is the amount of total work performed by maintenance.

    ☐ True

    ☐ False

2. Why should individual categories of the work load be defined?

    a. to establish proper size and craft composition of the work force

    b. to ensure clarity and common understanding of work load elements

    c. to avoid potential confusion

    d. all of the above

3. How are the following categories of work best controlled? Match the category in the left column (by letter) with the best control in the right column.

   a. PM services          _____ deliberately scheduled

   b. emergency repairs     _____ scheduled week by week

   c. planned maintenance   _____ immediate response

   d. unscheduled repairs   _____ fitted into existing work

4. Establishing the right number of personnel required by the work force is the essential purpose of work load determination.

   ☐ True

   ☐ False

5. Match the method for determining or limiting the labor requirement listed in left column (by letter) with the category of work listed in the right column.

   a. allocate labor                _____ emergency repairs

   b. do good preventive            _____ planned maintenance
      maintenance
                                    _____ routine maintenance
   c. perform jobs at regu-
      lar intervals                 _____ adjust work force when
                                          work load changes
   d. use backlog data

6. The organizational separation of mechanical and electrical crafts can:

   a. lead to an illusion of dealing with two separate maintenance departments

   b. create an illusion of little concern for labor control

   c. give the perception of little interest in labor control

   d. all of the above

7. Use the following words to complete the definition of "maintenance" in the answer blanks below: *existing, repair, condition, current.*

   Maintenance is the _________ and upkeep of _________ equipment, facilities, buildings, or areas in accordance with _________ design specifications to keep them in a safe, effective _________ while meeting their intended purposes.

8. Based on the definition of "maintenance," which of the following should be considered as nonmaintenance?

   a. relocation of equipment

   b. modification of equipment

   c. overhauls

   d. construction

   e. installation of new equipment

9. If a mechanic were to perform an inspection every 2 weeks and each service averaged 3 hours, what would be the annual work load for this service?

   a. 76 mechanical man-hours

   b. 78 mechanical man-hours

   c. 52 man-hours

   d. none of the above

10. If 30 minutes were added to each service described in the previous question for cleaning and to make minor adjustments, what would be the maximum work load?

    a. 76 hours

    b. 91 hours

    c. 59 hours

    d. none of the above

11. If a large unit of mobile equipment operated for an average of 120 hours per week, how many 250-hour services would be conducted in a 1-year period?

    a. 27

    b. 25

    c. 11

    d. none of the above

12. Which of the following best describes why emergency repairs may be needed?

    a. unexpected equipment stoppage during a scheduled operating period

    b. danger to personnel

    c. possible additional damage to equipment

    d. possible substantial production loss

    e. whenever operations deems necessary

13. Which of the following are characteristic of unscheduled repairs?

    a. it is nonemergency work of short duration

    b. it can be accomplished within a week

    c. there is little danger of equipment deterioration

    d. it is generally performed by one person

    e. it is usually completed in 2 hours or less

    f. it requires only shop stock or free-issue items

    g. it requires planning assistance

14. Which of the following are considered part of routine maintenance?

    a. craftsman training

    b. engine overhauls

    c. tool repair

    d. safety meetings

    e. all of the above

15. Selected major work is planned and scheduled because:

    a. it is a job of major scope

    b. of the high cost of job completion

    c. of the complexity of work

    d. of the importance of equipment

    e. all of the above

16. What are some of the advantages of planning and scheduling work?

    a. work is carried out more efficiently

    b. work is completed in the least amount of elapsed downtime

    c. planning and scheduling results in the most effective use of resources

    d. planning and scheduling gives the highest assurance of quality work

    e. all of the above

17. Scheduling of planned work is a joint maintenance–operations activity. What should be the department objectives at the scheduling meeting?

    a. to jointly agree on the time that work will interfere least with operations

    b. to enable maintenance to schedule their labor resources advantageously

    c. all of the above

18. Referring to the following maintenance labor utilization report, answer these questions:

    a. How many man-hours (MH) were spent completing unscheduled maintenance by electricians? _______

    b. What were the total MH spent on nonmaintenance by all crafts? _______

    c. What percentage of the total MH by all crafts were spent on emergency repairs? _______

**Maintenance Labor Utilization Report**
**Week 40—Ending September 30**
**Department 207**

| Craft | Preventive Maintenance | Scheduled Maintenance | Unscheduled Repairs | Emergency Repairs | Non-maintenance | Total Work Force |
|---|---|---|---|---|---|---|
| Millwright | 55 | 276 | 102 | 88 | 67 | 588 |
| Mechanic | 46 | 244 | 96 | 66 | 42 | 494 |
| Electrician | 23 | 122 | 54 | 32 | 46 | 277 |
| Welder | | 72 | 58 | 21 | 66 | 217 |
| Instrument | 18 | 32 | 30 | 12 | 45 | 137 |
| Pipefitter | | 48 | 12 | 9 | 12 | 81 |
| Laborer | | 121 | 6 | 5 | 41 | 173 |
| Total MH | 142 | 915 | 358 | 233 | 319 | 1,967 |
| % Distribution | 7 | 47 | 18 | 12 | 16 | 100 |

# CHAPTER 7

# Improving Work Force Productivity

1. What is the best way for maintenance to control the cost of the work they do?

    a. increase the efficiency with which they install materials

    b. use a smaller work force

    c. do work less often

    d. plan more work

2. Which of the following best describes labor productivity?

    a. percentage of time the work force is available for productive work during the shift

    b. percentage of time the work force is at the work site with tools performing productive work

3. Poor labor utilization is a factor of:

    a. bad work habits

    b. inadequate supervision

    c. poor attitudes

    d. all of the above

4. Poor labor productivity is linked to:

    a. poor preparation for work

    b. inadequate coordination

    c. not knowing what to do

    d. all of the above

5. What is the best method of improving labor productivity?

    a. implementing a well-defined and executed maintenance program

    b. implementing better discipline

    c. providing quality leadership

    d. measuring productivity

    e. all of the above

6. What are some of the factors that reduce labor utilization?

    a. late starts

    b. unsanctioned breaks

    c. sanctioned breaks

    d. in-plant travel

    e. early quits

    f. all of the above

7. What are some of the factors that reduce productivity?

    a. work not effectively planned

    b. poorly coordinated scheduling

    c. labor use not allocated properly

    d. materials and tools not ready

    e. all of the above

8.  What is meant by the term *random sampling*?

    a.  data gathering based on probability

    b.  observations at random intervals

    c.  providing a reliable picture of activities at all times

    d.  all of the above

9.  Random sampling is limited to observing:

    a.  work time

    b.  idleness

    c.  travel

    d.  clerical functions

    e.  time spent waiting

    f.  all of the above

10. What are the advantages of involving craft personnel directly in productivity measurements?

    a.  provides an opportunity for education

    b.  reduces resistance to being measured

    c.  reduces suspicion

    d.  gains support for necessary changes

    e.  all of the above

11. What are some of the reasons why productivity measurements might be resisted?

    a.  measurements could be misunderstood

    b.  work force might suspect layoffs

    c.  personnel may feel threatened

    d.  measurements might not be accurate

    e.  all of the above

12.  An industrial engineer can effectively measure five productivity-related factors: waiting, travel, clerical work, idleness, and work time. By involving craft personnel in measuring productivity, they can also evaluate many other factors that cause work delays, such as:

    a.  identifying parts

    b.  identifying broken tools

    c.  waiting for another craft

    d.  waiting for equipment to be shut down

    e.  using incorrect repair techniques

    f.  more training required

    g.  all of the above

13.  What beneficial impact could the involvement of craft personnel in measuring productivity have on the organization?

    a.  productivity measurements will be better understood

    b.  resistance to measurements will be reduced

    c.  suspicion will be overcome

    d.  better support of actions to improve productivity will result

    e.  all of the above

# CHAPTER 8

# Understanding Preventive Maintenance

1. What are the objectives of preventive maintenance?

    a. to extend equipment life

    b. to avoid premature equipment failure

    c. to conduct overhauls

2. What is characteristic of PM services?

    a. it is detection oriented

    b. it is done routinely (same checklist)

    c. it is done repetitively (repeated at regular intervals)

    d. all of the above

3. What does a static PM service mean?

    a. it requires a checklist

    b. it is performed when equipment is shut down

    c. it is performed by operations

4. Distinguish between a fixed and variable interval for the conduct of PM services.

   A fixed interval is _____.

   A variable interval is _____.

   a.   done every 2 weeks

   b.   done when conditions permit

   c.   done as needed

   d.   done every 250 operating hours

5. What are the advantages of detection-orientation in a PM program?

   a.   problems are found before equipment failure

   b.   problems will be less serious when found

   c.   it provides an opportunity to plan work

   d.   all of the above

6. Why will better preventive maintenance permit more planning?

   a.   planners run the PM program

   b.   it provides more time between discovery of a problem and actual repair

   c.   bigger jobs are found

   d.   none of the above

7. Identify the potential benefits of a successful PM program:

   a.   minimizes downtime

   b.   prolongs equipment life

   c.   could improve productivity

   d.   increases equipment reliability

   e.   increases opportunity for planned work

   f.   all of the above

8.  Mark the following activities with the appropriate type of service:

    R = routine PM services, or

    C = condition monitoring

    shock-pulse diagnosis ________

    cleaning ________

    testing and calibration ________

    ultrasonic testing ________

    inspection ________

    infrared scanning ________

9.  After a new PM program is started, what could happen if there were no competent planning activity to support it?

    a.  the same problems would be found again and again

    b.  deficiencies would not be converted into planned work

    c.  craftsmen might become discouraged

    d.  all of the above

10. How would a successful PM program change the labor used on:

| | Increase | Decrease |
|---|---|---|
| emergency repairs | ☐ | ☐ |
| planned maintenance | ☐ | ☐ |
| unscheduled repairs | ☐ | ☐ |

11.  Assign the following activities to the organizing versus operating steps of a PM program:

     a. modify checklists

     b. list equipment for PM

     c. balance labor needs

     d. set service intervals

     e. report results

     f. determine services required

     organizing steps:  ______  ______  ______

     operating steps:  ______  ______  ______

12.  When planned work, unscheduled repairs, or the need for emergency repairs are uncovered as a result of PM inspections, certain actions are required. Which of the actions listed below should be taken after discovery for each work category?

     Actions:

     a. apply planning criteria

     b. discuss with planner

     c. assign as soon as possible

     d. hold pending next shutdown

     planned work:  ______  ______  ______

     unscheduled repairs:  ______  ______  ______

     emergency repairs:  ______  ______  ______

13.  Based on the entire chapter, speculate on a logical order of events resulting from the effective use of a detection-oriented PM program. Number the events in the order they occur.

_____ planned work is completed in less elapsed downtime

_____ emergency work is avoided by early discovery of problems

_____ equipment is put back into production sooner

_____ planned work is done more deliberately with greater quality

_____ profitability is improved

_____ planned work is completed with less labor

_____ PM services find equipment problems in time to plan work

_____ resulting deficiencies can be converted into planned work

_____ planned work is completed at less overall cost

14.  What are some of the services performed by maintenance engineering?

a.   developing or modifying the PM program

b.   recommending nondestructive testing techniques

c.   developing standards for major jobs

d.   helping planners establish the scope of planned jobs

e.   recommending standard repair techniques

f.   all of the above

15.  Using the graph on page 43:

    a.  List the mechanical technologies you would consider for monitoring crushers.

    b.  List the electrical technologies you would consider for monitoring electric motors.

    c.  List the mechanical and electrical technologies you would consider for monitoring mobile equipment.

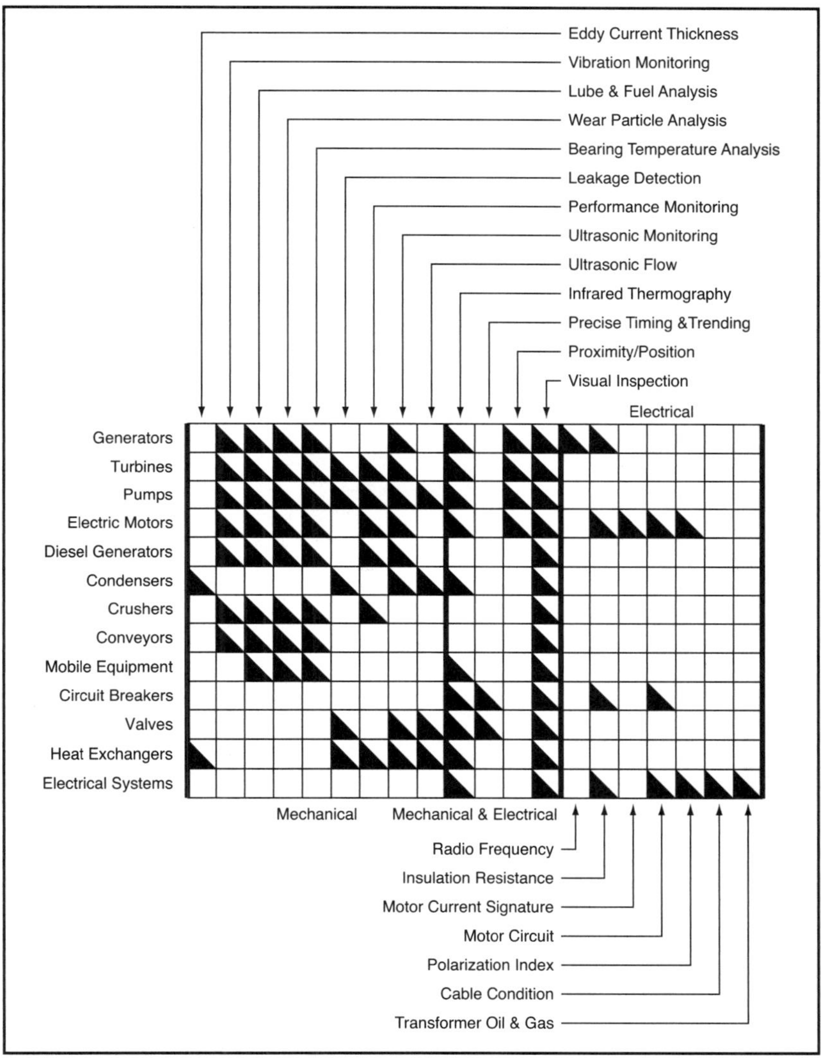
Eddy Current Thickness
Vibration Monitoring
Lube & Fuel Analysis
Wear Particle Analysis
Bearing Temperature Analysis
Leakage Detection
Performance Monitoring
Ultrasonic Monitoring
Ultrasonic Flow
Infrared Thermography
Precise Timing &Trending
Proximity/Position
Visual Inspection
Electrical
Generators
Turbines
Pumps
Electric Motors
Diesel Generators
Condensers
Crushers
Conveyors
Mobile Equipment
Circuit Breakers
Valves
Heat Exchangers
Electrical Systems
Mechanical
Mechanical & Electrical
Radio Frequency
Insulation Resistance
Motor Current Signature
Motor Circuit
Polarization Index
Cable Condition
Transformer Oil & Gas

# Effective Planning and Scheduling

1.  What distinguishes planning from scheduling? Place a "P" for planning or an "S" for scheduling after each action listed.

    a.  pre-organize selected major jobs ____

    b.  identify resources needed ____

    c.  ensure that work can be done in the least amount of downtime ____

    d.  ensure the most effective use of resources ____

    e.  complete the job at the lowest possible cost ____

    f.  determine the best time to complete the work ____

2.  How does the quality of preventive maintenance benefit planning?

    a.  detection-orientation preventive maintenance identifies problems before failure

    b.  problems are found far enough in advance to permit planning

    c.  all of the above

3.  Why are criteria necessary to determine what work should be planned?

    a.  to ensure that the planner is used effectively

    b.  to clarify working relationships between supervisors and planners

    c.  all of the above

4.  How does planning create an impact before work begins?

    a.  the job scope is properly identified

    b.  a solid job plan is organized

    c.  supervisors know what is expected

    d.  crews are advised of tasks in advance

    e.  materials can be preordered

    f.  all of the above

5.  When a job is planned, numerous potential benefits accrue. Identify these benefits:

    a.  less labor will be used

    b.  overall costs will be reduced

    c.  jobs will be completed with less elapsed downtime

    d.  better quality work will be produced

    e.  all of the above

6. Planning and scheduling steps are shown in the following list. Number them in the proper order.

    _____ conduct the scheduling meeting

    _____ determine if standards apply

    _____ open work order and order materials and shop work

    _____ confirm the job scope

    _____ make the job plan and set up the work order

    _____ confer with operations on job timing

    _____ identify the work

    _____ determine resources

    _____ establish labor by craft

    _____ investigate

    _____ estimate cost, set the job priority, and get approval

    _____ estimate preliminary time to do job

    _____ await receipt of materials

    _____ get advice from the crew

    _____ arrange for rigging, transport, and tools

    _____ monitor job execution and note cost and performance

7. A unique work order is required for every planned and scheduled job. What information does the maintenance planner record on the work order in order to manage the job from inception to completion?

    a. a work order number

    b. job identification

    c. resources required

    d. cost estimates

    e. task lists

    f. bills of materials

    g. tool lists

    h. all of the above

8. What are some of the objectives of the joint operations and maintenance scheduling meeting?

    a. agree to perform work when it interferes least with operations

    b. make the best use of maintenance resources

    c. obtain an approved schedule

    d. review and improve the previous week's schedule compliance

    e. all of the above

9. At what point can materials, shop work, or labor can be charged against the work order?

    a. any time during the planning process

    b. only after the work order is opened

    c. as soon as the shop is ready to do the work

    d. whenever the warehouse has the parts

    e. none of the above

10. What information does the planner utilize to manage each planned job?

    a. accumulated material costs

    b. actual versus estimated labor hours

    c. labor cost

    d. material cost

    e. job cost and performance

    f. all of the above

11. What are some of the information sources the planner can use to develop new planned and scheduled work?

    a.  cost reports to help identify costly equipment

    b.  repair history to spot the chronic, repetitive problems

    c.  repair history to identify failure trends

    d.  field investigations about troublesome equipment

    e.  none of the above

12. What are some of the desirable characteristics of an effective maintenance and operations scheduling meeting?

    a.  it is attended by principal supervisors

    b.  binding decisions can be made

    c.  operations commits to making equipment available

    d.  maintenance identifies equipment to maintain and repair duration

    e.  maintenance states the risk if equipment is not made available

    f.  compliance and performance can be checked for the current week

    g.  attendance includes engineering, material control, and safety

    h.  plant manager should attend

    i.  it should be supplemented with daily coordination meetings to adjust for unexpected delays

    j.  all of the above

13. What steps should the planner take before placing a work order on the proposed schedule?

    a.  identify jobs that should be done the following week

    b.  verify that materials will be available

    c.  confirm completion of shop work

    d.  ensure that the labor resources are available

    e.  all of the above

14. The scheduling process is divided into five phases:

    1. getting ready
    2. conducting the scheduling meeting
    3. advising key personnel
    4. checking compliance and performance
    5. following up during schedule execution

    The steps carried out for each phase are randomly listed. Place a logical phase number (1–5) beside each step.

    _____ distribute the approved schedule

    _____ allocate labor

    _____ check performance indices

    _____ confirm priorities

    _____ explain the tasks

    _____ prepare preliminary plan

    _____ confirm shutdown times

    _____ determine completeness of planning

    _____ confirm use of equipment

    _____ discuss within maintenance

    _____ negotiate with operations

    _____ confirm delivery of materials

    _____ verify availability of materials

    _____ visit field location to observe work in progress or detect developing problems

    _____ obtain schedule approval

    _____ verify completion of shop work

    _____ check schedule compliance

    _____ propose shutdown times

    _____ verify work force

15. How should static PM services versus dynamic PM services be scheduled?

    a. static PM services require equipment shutdown

    b. dynamic PM services are done with equipment running

    c. all PM services must be scheduled by planners

16. Forecasting is the identification of the most likely future time when major components such as engines, drive lines, or pumps may have to be replaced. Select the precautions necessary when replacing major components:

    a. verify component condition before replacing

    b. carefully check recent repair history

    c. establish the cause of failure

    d. all of the above

17. What is meant by the phrase "mean time before failure (MTBF)"?

    a. the average life-span of a range of similar components

    b. the time necessary to make repairs

    c. none of the above

18. Typically, a maintenance standard is developed using historical information from previous similar jobs concerning what was done and how, the materials used, tools required, and labor utilized. Standards have direct applicability to forecasting and their use can aid the planning task. How?

    a. planners can plan more jobs

    b. the planning task is simplified

    c. job performance can be accurately measured

    d. an opportunity to improve performance exists

    e. all of the above

# CHAPTER 10

# Reliability Centered Maintenance

1.  RCM features:

    a.  identification of specific equipment functions

    b.  assignment of equipment performance standards

    c.  identification of the nature of equipment failures

    d.  definition of when performance standards are not met

    e.  addition of significant condition-monitoring to the maintenance program

    f.  all of the above

2.  Strategies such as RCM augment equipment management because:

    a.  modern equipment is designed for greater reliability

    b.  equipment has greater productive capacity

    c.  equipment is more complex

    d.  equipment is increasingly challenging to maintain

    e.  greater technical expertise is required to realize potential

    f.  all of the above

3. Reliability centered maintenance (RCM) is a proactive strategy that challenges many of the practices of traditional maintenance, among them:

    a. time-based actions prove to be inadequate

    b. equipment reliability deteriorates with age

    c. all of the above

4. An RCM performance standard should include:

    a. functions of the equipment

    b. expectations of capacity, speed, and so forth

    c. circumstances of performance

    d. all of the above

5. Which of the following conditions constitute a functional failure?

    a. an identifiable physical condition which indicates that the failure process has started

    b. the inability of a unit of equipment or component to meet a specified performance standard

6. Which of the following conditions are characteristic of equipment failure?

    a. equipment user determines if failure has occurred

    b. deviation from the normal operating condition

    c. unsatisfactory condition as viewed by equipment user

    d. may range from possible stoppage to a complete equipment failure

    e. all of the above

7. Place a "P" or an "F" after the condition to indicate which is a potential (P) or a functional (F) failure:

    a. doesn't perform functions to performance standard specified ____

    b. an indication that the failure process has started ____

8. What are the functions of condition-monitoring techniques as applied in RCM?

    a. early detection of potential failures

    b. distinguish failures from normal operating conditions

    c. forestall failure and its consequences

    d. all of the above

9. Condition-monitoring techniques used to detect potential failures are called "on-condition" tasks. What does this mean?

    a. equipment is left in service if it meets performance standard

    b. the equipment runs

    c. none of the above

10. What is meant by the P–F interval?

    a. the interval between potential failure and functional failure

    b. the mean time before failure

    c. a list of on-condition tasks

    d. all of the above

11. Modern condition-monitoring techniques:

    a. allow potential failures to be detected sooner

    b. yield a longer P–F interval

    c. provide more time to take corrective actions

    d. help avoid functional failure and its consequences

    e. all of the above

12. Implementing RCM is an eight-step process. Number the steps listed in the proper order:

    _____ identify the functions of the equipment

    _____ determine the types of failures

    _____ enumerate the consequences of failures

    _____ select the most critical equipment

    _____ rank the consequences of failures

    _____ incorporate RCM into the overall maintenance plan

    _____ establish performance standards

    _____ apply the most effective condition-monitoring techniques

13. What are some sensible guidelines for identifying critical equipment?

    a. involve all departments

    b. use a simple formula

    c. do not get bogged down in details

    d. listing can be changed if warranted

    e. get implementation under way quickly

    f. subsequent experience may alter listing

    g. all of the above

14. Which of the following conditions best describe equipment reliability?

    a. operating hours until failure

    b. the extended life-span of internal components

    c. none of the above

15. Before equipment failure patterns were well understood, it was commonly believed that:

    a. equipment wore out as it got older

    b. equipment required breaking in

    c. equipment had a life of its own

16. Prior to the introduction of RCM, prevailing solutions for avoiding failures included replacing major components and overhauling equipment after a predetermined wear-out period.

    ☐ True

    ☐ False

17. What is the primary means by which data are developed to determine the nature of failure patterns and the degree to which failures occur versus the life-span of the equipment or components?

    a. review repair history

    b. review cost data

    c. go by personal experience

    d. trial and error

18. Match failure patterns in the left column with potential corrective measures or causes in the right column.

| | |
|---|---|
| failure to detect problem _____ | a. poor selection of repair material |
| operating error _____ | |
| material fatigue _____ | b. more monitoring needed |
| infant mortality _____ | c. poor supervision or workmanship |
| random failures _____ | d. improper operating procedures |
| | e. inconsistent maintenance practices |

19. Random failure patterns can occur at any time. To guard against the unexpected, RCM must include which of the following responses:

    a. reactive

    b. time based

    c. condition based

    d. proactive

    e. all of the above

20. The P–F interval is the elapsed time between the detection of a potential failure (P) until a functional failure (F) will occur if corrective actions are not taken. Which of the following conditions best describe the determination of the P–F interval and actions that are taken after it has been established?

    a. condition monitoring identifies P

    b. repair history provides interval until F

    c. a longer P–F interval extends time for corrective actions

    d. during this period, equipment is more carefully monitored

    e. if left in operation, the equipment can perform designated function

    f. actions during this period try to restore normal operation

    g. if successful, equipment life-span is extended

    h. all of the above

21. How might the nature of the PM program "detection" processes be changed with RCM implementation?

    a. time-based inspections may change to condition-based

    b. time-based inspections might be expanded with condition-based inspections

    c. some equipment might be subject to redesign

    d. noncritical equipment might be run to failure

    e. all of the above

22. If equipment displays random failure patterns at any point in its repair history, condition-based monitoring alone or in combination with time-based inspections will be more effective in reducing the instances of failure.

    ☐ True

    ☐ False

23. Repair history records the chronological listing of significant repairs made on critical equipment. Specifically, what does this include?

    a. the nature of chronic, repetitive problems

    b. confirmation of failure modes

    c. observation of failure trends

    d. determination of failure frequencies

    e. the life-span of components

    f. determination of the adequacy of equipment monitoring

    g. determination of whether corrective actions have been effective

    h. all of the above

24. After the P–F interval has been established, an on-condition status is assigned to the equipment being observed. Observation includes more vigorous monitoring to establish the underlying nature of the newly discovered potential failure. During this period, useful maintenance, operations, or material solutions are actively sought to try to restore the equipment to its normal operating condition.

    ☐ True

    ☐ False

25. What are the most common proactive reliability tools used in RCM to establish the underlying causes of failure and to assess associated risks?

    a. root cause failure analysis (RCFA)

    b. failure modes and effects analysis (FMEA)

    c. risk-based inspections (RBIs)

    d. trial and error

26. What are the logical steps involved in the investigation of equipment failures?

    a. inspection of the damaged equipment

    b. discussions with involved personnel

    c. examination of pertinent records

    d. assessment of contributing factors

    e. none of the above

27. Distinguish between RCFA, FMEA, and RBI by matching the tool (left column) with the correct statement (right column).

    RCFA _____          a. studies failure causes and ranks
                           their risk
    FMEA _____
                        b. principal application is piping
    RBI _____
                        c. determines why failure happened
                           and corrective actions

28. Mark each of the following statements as true (T) or false (F).

    _____ total cost of reliability is the cost of failures not averted

    _____ failures result in injuries, lost product, and equipment damage

    _____ maintenance programs act to reduce the cost of maintenance

    _____ successful RCM attacks the cost of failure

    _____ failure avoidance increases reliability

# CHAPTER 11

# Total Productive Maintenance

1.  Total productive maintenance (TPM) brings together the American practice of preventive maintenance with the Japanese concept of total quality control. What are its objectives?

    a.  to maximize equipment effectiveness

    b.  to establish preventive maintenance for the life-span of equipment

    c.  to involve all departments in a joint effort

    d.  to gain support of every plant employee

    e.  to promote itself through small-group activities

    f.  all of the above

2.  What are some of the tasks that operators can perform competently?

    a.  cleaning

    b.  inspecting

    c.  adjusting

    d.  calibration

    e.  replacing major components

3.  TPM includes productive maintenance. What is included in productive maintenance?

    a.  equipment design to minimize maintenance

    b.  emphasis on preventive maintenance to preclude breakdowns

    c.  modification to make maintenance easier

    d.  equipment replacement policies

4.  Autonomous maintenance is the unique aspect of TPM. What does it include?

    a.  maintenance work carried out in small groups

    b.  work such as adjustment, calibration, or lubrication

    c.  maintenance work carried out by operators

    d.  all of the above

5.  What does the word "total" mean in TPM?

    a.  efficiency and total plant operations

    b.  total avoidance of failures

    c.  total effort to conduct better maintenance

    d.  total effort to improve maintainability

    e.  total plant participation

    f.  all of the above

6.  Equipment automation causes a shift of maintenance emphasis from workers to machines. What does this suggest?

    a.  output is affected more by equipment condition

    b.  output is affected more by the efforts of workers

    c.  automation will produce more product

    d.  automation reduces human error

    e.  automation is cheaper

    f.  none of the above

7. Overall equipment effectiveness avoids six big losses. What are they?

    a. equipment failure

    b. lost time for setup and adjustments

    c. idle equipment and minor stoppages

    d. reduced equipment speed

    e. process defects

    f. reduced equipment yields

    g. maintenance downtime

    h. none of the above

8. TPM suggests that the traditional pattern of failure starts with early failures followed by a random failure period and ending with a wear-out period. What solutions are proposed for better maintenance to counter early failures and extend the wear-out period?

    a. better preventive maintenance will reduce early failures

    b. better maintenance can extend the wear-out period

    c. random failures simply occur

    d. condition monitoring can reduce random failures

    e. none of the above

9. Five countermeasures are emphasized to help reduce or eliminate failures. What are they?

    a. keep equipment in good condition

    b. operate equipment properly

    c. restore equipment with effective maintenance

    d. consider design changes to minimize maintenance

    e. improve people skills through training

    f. reorganize

10. Life-cycle costing is used to determine which maintenance procedures are less costly and more effectively performed in a certain way. Consider the following chart.

| Paint Type | Materials, $ | Labor, $ | Life-span, years | Costs, $ 3 years | Costs, $ 6 years |
|---|---|---|---|---|---|
| A | 5,000 | 20,000 | 3 | 25,000 | 50,000 |
| B | 15,000 | 20,000 | 6 | | 35,000 |

Paint A would last for 3 years and materials would cost $5,000. It would require $20,000 in labor to paint the designated area. Paint B would cost $15,000 and also require $20,000 in labor but it would last for 6 years. At the end of the first 3 years, paint A must be reapplied, incurring an additional $25,000 expenditure for the next 3 years.

Which paint would be the most cost-efficient:

☐ Paint A

☐ Paint B

11. Number the following steps in logical order to ensure the effective implementation of TPM:

_____ initiate autonomous maintenance

_____ increase skills of operators and maintainers

_____ reduce the six big losses

_____ apply these factors to specific equipment

12. TPM implementation is carried out in three stages. Match the stages listed in the left column with the related activities in the right column.

preparatory _____          a. normalize the total operation

implementation _____          b. prepare personnel, establish goals

stabilization _____          c. establish the TPM program

# CHAPTER 12

# Implementing Information Systems

1. Education of system users is a key factor in successful information system implementation. What are some of the key factors in this education process?

    a. associating duties with information needed

    b. emphasizing information needed to perform duties

    c. training team members together

    d. emphasizing supervisors' work control responsibilities

    e. all of the above

2. What tasks should supervisor training include?

    a. review of outstanding jobs

    b. determine dynamic PM services to be done

    c. make decisions on jobs to be done

    d. assign jobs to crew members

    e. provide job instructions

    f. monitor job status

    g. close out completed, unplanned jobs

    h. all of the above

3. Typically, which of the following should the planners' training *not* include?

   a. use of work orders for planned jobs

   b. determination of work order status

   c. control of the supervisor's crew

   d. determination of work order cost

   e. observation of the backlog

   f. observation of the use of labor

4. If a PM service requires no equipment shutdown and can be accomplished anytime within the week it is scheduled, who is in the best position to schedule it?

   a. a crew member

   b. the planner

   c. the supervisor

   d. operations

5. What are some of the logical aspects of information that should be included in training for craftsmen?

   a. entry and use of equipment specifications data

   b. initial reporting of labor data (verified by supervisor)

   c. updating of equipment specifications data

   d. review of repair history data

   e. entering of new work

   f. all of the above

6. Why is computer literacy less of a problem today among craftsmen?

   a. planners supply all needed information

   b. there is an increase in the use of personal computers

   c. craftsmen do not need information

   d. none of the above

7. Why is it best to include craftsmen in the initial reporting of field data such as labor use or repair history?

    a. supervisors can verify initial data reported

    b. craftsmen can report accurately because they did the actual work

    c. none of the above

8. Successful implementation of an information system requires that specific steps be followed. Number the following randomly listed steps in logical order.

    _____ load all files and confirm equipment numbering

    _____ verify that hardware and networking arrangements are functional

    _____ conduct effective, realistic training

    _____ verify field data source accuracy

    _____ ensure that roles of key personnel are correct and understood

    _____ phase out all previous, conflicting procedures

    _____ establish a core training group

    _____ confirm the soundness of the equipment management program

    _____ develop an implementation schedule with specific objectives

    _____ monitor system use and accomplishments

9. The work order system is exclusively for the use of maintenance.

    ☐ True

    ☐ False

10. Which of the following functions are *not* related to work order use?

    a.  applying to all types of work performed

    b.  documenting the job cost and performance

    c.  allowing the accumulation of costs to jobs, equipment, or functions

    d.  providing for the control of routine, repetitive functions

    e.  allowing for the use of verbal orders (with followup)

    f.  directly reporting labor and material use

    g.  providing simple controls for actions such as shop cleanup

    h.  providing controls suitable to the work being done

    i.  linking production statistics to yield performance information

    j.  using accounting to yield resource use information

11. Work orders allow maintenance to carry out which of the following functions?

    a.  ordering service

    b.  identifying work required

    c.  relating work to a unit of equipment

    d.  identifying functions to be performed

    e.  establishing the type of work, timing, and importance

    f.  identifying labor, material, and equipment requirements

    g.  making provisions for job approval

    h.  connecting field data to accounting

    i.  facilitating the use of standards

    j.  providing data for the accounting system

    k.  serving as the basis for scheduling work

    l.  all of the above

12. Match the category of work performed in the right column with the work order element that would normally support it in the left column.

| | |
|---|---|
| maintenance work order (MWO) ____ | a. planned and scheduled work |
| maintenance work request (MWR) ____ | b. running repairs |
| verbal orders ____ | c. initiate emergency repairs |
| standing work order (SWO) ____ | d. routine, repetitive actions |
| engineering work order (EWO) ____ | e. nonmaintenance work |

13. What functions does the work order number allow when the work order is opened and data flows through the system to yield information?

    a. it establishes a link with equipment on which work will be done

    b. it relates job details to hours used and materials consumed

    c. it isolates cost and performance of a job

    d. it allows specific job assignments to crew members

    e. it links MH charged to job by crew

    f. it accumulates labor costs and MH against the job

    g. it compares labor cost and actual MH with estimates

    h. it compares job cost with the estimate

    i. all of the above

14. Elements of the work order system, the accounting system, and production control statistics function in concert to develop information. Identify which elements listed belong to the work order system (W), the accounting system (A), or are production data (P).

    _____ time card

    _____ SWO

    _____ tons of throughput

    _____ stock issue card

    _____ EWO

    _____ purchase order

    _____ MWO

    _____ operating hours

15. A logical and consistent numbering scheme should be applied to fixed and mobile equipment as well as to functions. What should the numbering scheme include?

    a. department or cost center

    b. type of equipment or a function

    c. equipment components

    d. subdivisions of functions (e.g., PM inspection, route 3)

    e. all of the above

16. What is the purpose of failure coding?

    a. to identify the cause of failure for individual jobs

    b. to reveal a pattern of failures over an extended period

    c. to dictate specific corrective actions based on results

    d. all of the above

17.  A standard work description provides certain advantages, among them:

    a.  it provides an additional file search

    b.  it allows development of detail for standardizing jobs

    c.  it speeds the visual search of repair history

    d.  all of the above

18.  Which of the following best describes the objectives of priority-setting?

    a.  to avoid the first-in/first-out syndrome

    b.  to identify the importance of a job compared with all others

    c.  to help to avoid arguments

    d.  to establish a time period within which the job should be completed

    e.  none of the above

19.  An approved schedule issued after an operations–maintenance scheduling meeting contains static PM services and major jobs. On receipt of the schedule, what logical actions might be taken by the maintenance supervisor whose crew will carry out the work?

    a.  add unscheduled jobs on equipment being shut down for major jobs

    b.  add work requests on equipment undergoing static PM services

    c.  prepare daily and shift crew assignments

    d.  all of the above

# CHAPTER 13

# Essential Information

1. Different maintenance organization levels require specific information that will permit them to make informed decisions. What are some of the decision-making information needs of maintenance managers?

   a. a monthly absentee summary

   b. compliance with the vacation policy

   c. total maintenance costs year-to-date

   d. forecast of major component replacements for 6 months

2. Match the information needs in the right column with the logical user in the left column.

   maintenance engineers _____

   general supervisor _____

   planner _____

   supervisor _____

   craftsmen _____

   a. cost and performance of a major job

   b. cost to replace a major component

   c. percentage completion of an overhaul

   d. list of interchangeable bearings

   e. life-span of engines in fleet A

3.  What characterizes decision-making information?

    a.  manage the overall maintenance effort

    b.  manage internal maintenance functions

    c.  none of the above

4.  Speculate on the most logical objectives of maintenance information listed below:

    a.  to determine what work must be done

    b.  to justify actions required

    c.  to measure the effectiveness of work done

    d.  to give the planner job security

5.  Labor utilization information is vital because:

    a.  the work force exists to install materials

    b.  labor use reduces downtime

    c.  work cost depends on material installation efficiency

    d.  less downtime means greater reliability

6.  Referring to the maintenance labor utilization report on page 75, answer the following questions:

    a.  How many MH did mechanics use on scheduled maintenance? _______

    b.  How many MH were used on emergency repairs? _______

    c.  How many electrician MH were used during the week? _______

    d.  What percentage of the work done was nonmaintenance? _______

    e.  How many total MH were used? _______

**Maintenance Labor Utilization Report**
**Week 40—Ending September 30**
**Department 207**

| Craft | Preventive Maintenance | Scheduled Maintenance | Unscheduled Repairs | Emergency Repairs | Non-maintenance | Total Work Force |
|---|---|---|---|---|---|---|
| Millwright | 55 | 276 | 102 | 88 | 67 | 588 |
| Mechanic | 46 | 244 | 96 | 66 | 42 | 494 |
| Electrician | 23 | 122 | 54 | 32 | 46 | 277 |
| Welder | | 72 | 58 | 21 | 66 | 217 |
| Instrument | 18 | 32 | 30 | 12 | 45 | 137 |
| Pipefitter | | 48 | 12 | 9 | 12 | 81 |
| Laborer | | 121 | 6 | 5 | 41 | 173 |
| Total MH | 142 | 915 | 358 | 233 | 319 | 1,967 |
| % Distribution | 7 | 47 | 18 | 12 | 16 | 100 |

7.  Which categories of work are *not* in the backlog?

    a.  preventive maintenance

    b.  emergency repairs

    c.  unscheduled repairs

    d.  planned/scheduled maintenance

8.  Referring to the backlog summary report provided below, answer the following questions.

   a.  In man-hours estimated (MHE), what was the change in backlog hours for welders? _______

   b.  Which craft had the biggest increase in the backlog?

   _______________________________________

   c.  How many MHE were added for electricians? _______

   d.  Which craft had the greatest backlog reduction?

   _______________________________________

   e.  How many total MHE were added for the week? _______

   f.  What was the total change in the backlog for the week?

   _______

| | | | Backlog Summary Report<br>Week 40—Ending September 30<br>Maintenance Backlog | | |
|---|---|---|---|---|---|
| **Craft** | **ST WK** | **+ WK** | **– WK** | **ND WK** | **± BL** |
| MECH | 556 | 67 | 52 | 571 | +15 |
| ELEC | 427 | 51 | 82 | 396 | –31 |
| WELD | 336 | 96 | 36 | 396 | +60 |
| RIGR | 442 | 212 | 119 | 535 | +93 |
| FITR | 484 | 96 | 134 | 446 | –38 |
| HLPR | 314 | 201 | 187 | 328 | +14 |
| TOTAL | 2,559 | 723 | 610 | 2,672 | +113 |

9. *Backlog Computation.* Three relationships exist between the estimated and actual MH as work orders make their way through the development of backlog information:

   - actual MH are more than MHE

   - MHE are more than actual MH

   - actual MH are equal to MHE

Backlog computation rules are:

   - if actual MH are more than estimated, the backlog equals 0

   - if MHE are more than actual hours, cancel remaining estimated hours

   - there is no negative backlog

Consider the diagram on page 78. Nine jobs (work orders) are shown (A–I). Each job shows the MHE by craft (MECH, MW, ELEC). The single heavy horizontal line depicts the week that the work order was opened and the MHE for each craft were added to the backlog. These MH remain in the backlog until the work is completed and the work order is closed. The double heavy horizontal line signals the week when work was started and completed as well as the amount of actual MH used by craft. The backlog data graph under the diagram shows the number of backlogged MH for each craft at the start of week number 10. Determine and fill in the correct backlog by craft and the total backlog at the end of week 18.

| | | | 10 | 11 | 12 | 13 | 14 | 15 | 16 | 17 | 18 | 19 |
|---|---|---|---|---|---|---|---|---|---|---|---|---|
| **A** | MECH | 150 | 150 | | | | 150 | 0 * | | | | |
| | MW | 70 | 70 | | | | 60 | 10 | | | | |
| | ELEC | 25 | 25 | | | | 35 | 0 | | | | |
| **B** | MECH | 70 | 70 | | | 90 | | 0 * | | | | |
| | MW | 20 | 20 | | | 10 | | 10 | | | | |
| | ELEC | 15 | 15 | | | 25 | | 0 | | | | |
| **C** | MECH | 210 | | 210 | | | 180 | 30 * | | | | |
| | MW | 40 | | 40 | | | 45 | 0 | | | | |
| | ELEC | 30 | | 30 | | | 20 | 10 | | | | |
| **D** | MECH | 140 | | 140 | | | 105 | 50 | 0 * | | | |
| | MW | 75 | | 75 | | | 50 | 25 | 0 | | | |
| | ELEC | 10 | | 10 | | | 10 | 5 | 0 | | | |
| **E** | MECH | 110 | | | 110 | | | | 90 | 20 * | | |
| | MW | 45 | | | 45 | | | | 60 | 0 | | |
| | ELEC | 15 | | | 15 | | | | 15 | 0 | | |
| **F** | MECH | 145 | | | | 145 | | | | | 135 | 10 * |
| | MW | 85 | | | | 85 | | | | | 70 | 15 |
| | ELEC | 10 | | | | 10 | | | | | 15 | 0 |
| **G** | MECH | 150 | | | | | 150 | | | | 175 | 0 * |
| | MW | 70 | | | | | 70 | | | | 80 | 0 |
| | ELEC | 25 | | | | | 25 | | | | 20 | 5 |
| **H** | MECH | 180 | | | | | 180 | | | | 110 | 35 |
| | MW | 60 | | | | | 60 | | | | 70 | 10 |
| | ELEC | 30 | | | | | 30 | | | | 20 | 20 |
| **I** | MECH | 145 | | | | | | 145 | | | | 110 |
| | MW | 60 | | | | | | 60 | | | | 60 |
| | ELEC | 15 | | | | | | 15 | | | | 25 |

| Backlog Data | | | | | | | | | | | |
|---|---|---|---|---|---|---|---|---|---|---|---|
| Mechanic | 1,270 | | | | | | | | | | |
| Millwright | 1,455 | | | | | | | | | | |
| Electrician | 710 | | | | | | | | | | |
| Total | 3,435 | | | | | | | | | | |

10. Consider the following backlog diagram. Note that the crafts are numbered "1," "2," and "3," and the backlog has increased during the 8-week period illustrated.

    a.  Which craft may have too many personnel? ______

    b.  Which craft seems to have the right number of personnel? ______

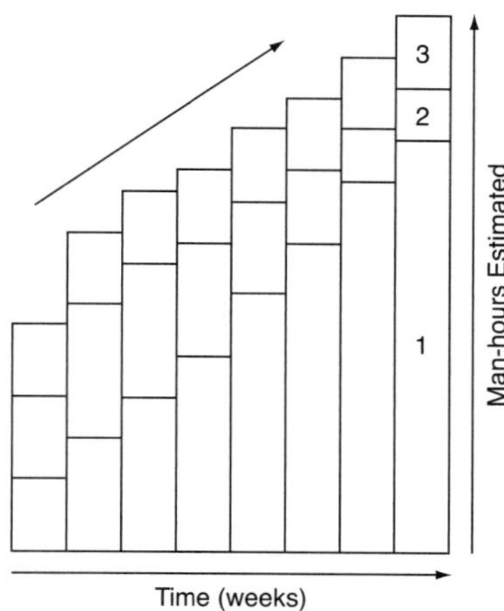

11. Referring to the maintenance work order status report provided below, answer the following questions:

    a. What was the actual accumulated labor cost? __________

    b. What was the estimated cost of purchased materials?

    __________

    c. How many MH were estimated for craft 1? __________

    d. When was this work order opened? __________

    __________

    e. What was the serial number of the new drive motor?

    __________

    f. What was the job priority? __________

    __________

---

**Maintenance Work Order Status Report**

| MWO# | Section | Type | Unit | Component | Category |
|---|---|---|---|---|---|
| 514706 | 07 | 05 | 011 | 03 | Scheduled |

Description of Work: RPL Drive Motor

| Status | Open | Start | Priority | WA | FC | SN/In | SN/Out |
|---|---|---|---|---|---|---|---|
| Complete | 0719 | 0722 | 90 | NO | 8 | 34338A | 665470 |

| | Man-Hours | |
|---|---|---|
| | EST | ACT |
| Craft 1 | 125 | 141 |
| Craft 2 | 60 | 92 |
| Labor | 2,775 | 3,495 |
| Stock | 3,710 | 4,670 |
| Purchased | 2,100 | 2,210 |
| TOTAL | 8,585 | 10,375 |

Work Order Notes

See print 07-A14

| Requested: | Approved: | Completed: |
|---|---|---|
| RL Jones | PD Williamson | LF Brown |
| 04/03/07 | 04/16/07 | 04/22/07 |

12. Referring to the equipment cost by type/component report provided below, answer the following questions:

    a. Which component had the greatest average cost per month?

    _______________

    b. What was the total maintenance cost for February?

    _______________

    c. How many MH were used on frames through February? _______________

    d. What was the average cost per month for hydraulics?

    _______________

    e. What is wrong with the way the components are numbered? _______________________________

**Equipment Cost by Type/Component**
**February**

| | CURRENT MONTH | | | | | YEAR TO DATE | | | | | |
| | MH | | | | | MH | | | | | |
| TYPE CP | RT | OT | LBR CST | MTL CST | TTL CST | RT | OT | LBR CST | MTL CST | TTL CST | AVG CST MONTH |
|---|---|---|---|---|---|---|---|---|---|---|---|
| 51 LHDS-PRODUCTION | | | | | | | | | | | |
| 00 GENERAL | 1,021.5 | 29.0 | 10,666 | 10,895 | 21,571 | 2,595.0 | 150.5 | 28,933 | 25,756 | 54,697 | 27,344 |
| 01 ENGINE | 490.5 | 39.0 | 5,930 | 30,993 | 36,936 | 1,119.0 | 84.5 | 13,546 | 52,324 | 65,884 | 32,933 |
| 02 CONVERTER | 20.0 | 1.0 | 217 | 24 | 241 | 66.0 | 3.0 | 755 | 295 | 1,050 | 525 |
| 03 TRANS | 358.5 | 24.5 | 4,198 | 12,732 | 16,932 | 797.5 | 55.0 | 9,312 | 12,892 | 22,217 | 11,102 |
| 04 DRIVE TRAIN | 592.0 | 28.5 | 6,668 | 17,294 | 23,968 | 1,291.5 | 87.0 | 15,131 | 29,702 | 44,841 | 22,416 |
| 05 TIRES | 132.5 | 11.5 | 1,543 | 10,634 | 12,187 | 289.5 | 26.5 | 3,432 | 58,180 | 61,622 | 30,806 |
| 06 BUCKET | 479.5 | 23.0 | 5,504 | 2,800 | 8,310 | 770.5 | 30.5 | 8,715 | 8,389 | 17,110 | 8,552 |
| 07 HYDRAULICS | 625.5 | 44.5 | 7,331 | 7,166 | 14,509 | 1,664.0 | 102.0 | 19,438 | 11,477 | 30,929 | 15,457 |
| 08 BOOM | 43.0 | 0.0 | 485 | 1,485 | 1,946 | 200.0 | 3.5 | 2,306 | 4,002 | 6,312 | 3,154 |
| 09 BRAKES | 495.0 | 39.5 | 5,939 | 4,460 | 10,411 | 913.5 | 71.5 | 11,014 | 7,330 | 18,358 | 9,172 |
| 10 FRAME | 902.0 | 60.0 | 10,448 | 10,292 | 20,749 | 1,814.5 | 105.5 | 21,091 | 15,099 | 36,204 | 18,096 |
| 12 ELECTRICAL | 249.5 | 15.0 | 2,945 | 3,172 | 6,125 | 539.0 | 31.5 | 6,384 | 8,555 | 14,949 | 7,469 |
| 13 SERVICE | 10.5 | 0.0 | 112 | 0 | 112 | 38.5 | 0.0 | 405 | 0 | 405 | 132 |
| TOTAL | 6,411.0 | 315.5 | 61,996 | 111,920 | 173,997 | 12,096.5 | 751.0 | 140,462 | 234,001 | 374,578 | 187,210 |

13. Referring to the repair history report provided below, answer the following questions:

    a. How many operating hours (OpHrs) did SN 590234 last?

    _______________

    b. How many weeks did SN 112123 last? _______________

    c. What was the average number of MH (rounded up) used replacing these components? _______________

    d. What MWO number covered the replacement of SN 129032? _______________

    e. Which failure code (FC) is highest? _______________

| | | | | | | | | Repair History | | | |
|---|---|---|---|---|---|---|---|---|---|---|---|
| FC | MWO # | Sec | Unit | Comp | Tr | MH | YrWk | Description of Work | SN (In) | SN (Out) | OpHrs |
| 1 | 64236 | 024 | CM02 | 400 | EL | 8 | 0643 | RPL TRAM MTR OFF SIDE (JOY) | 112123 | 312413 | 1102 |
| 1 | 65135 | 024 | CM02 | 400 | EL | 6 | 0701 | RPL TRAM MTR OFF SIDE (JOY) | 223335 | 112123 | 1234 |
| 2 | 65223 | 024 | CM02 | 400 | EL | 8 | 0705 | RPL TRAM MTR OFF SIDE (JOY) | 124498 | 223335 | 1286 |
| 2 | 65465 | 024 | CM02 | 400 | EL | 5 | 0729 | RPL TRAM MTR OFF SIDE (TRAM) | 344456 | 124456 | 1589 |
| 2 | 65553 | 024 | CM02 | 400 | EL | 5 | 0708 | RPL TRAM MTR OFF SIDE (TRAM) | 129032 | 344456 | 1703 |
| 2 | 65664 | 024 | CM02 | 400 | EL | 6 | 0703 | RPL TRAM MTR OPP SIDE (JOY) | 590234 | 129032 | 3307 |
| 1 | 65701 | 024 | CM02 | 400 | EL | 8 | 0708 | RPL TRAM MTR OPP SIDE (NAT) | 452903 | 590234 | 3490 |

14. What is the ultimate objective of good maintenance performance?

    a. reduced downtime

    b. less maintenance cost

    c. profitability

    d. improved reliability

15. Which elements of information are most needed by: (1) management, (2) operations, and (3) maintenance? Place a 1, 2, or 3 in the spaces provided to match the area with the information.

_____ assurance that funds are well spent and quality work is done

_____ trends and summaries to assess maintenance effectiveness

_____ identify work, control work, and measure accomplishments

16. The following popular performance indices are associated with a specific interpretation:

a. tons per dollar

b. cost per operating hour

c. labor cost to install each dollar of material

Associate the each indices with its common interpretation given in the following list:

_____ how well maintenance contributes to reduction of downtime

_____ overall maintenance contribution to plant productive effort

_____ reveals how effectively maintenance uses its labor to perform work

17. An unsatisfactory performance revealed by a declining trend must be followed by a detailed investigation to identify problems and make corrections. What steps should management take to ensure improvement?

a. seek explanation from maintenance

b. require followup corrective actions

c. continue to observe indices

d. note impact of corrective actions

e. all of the above

18. What are some additional performance indices management should require of maintenance?

    a.  compliance with the PM schedule

    b.  compliance with the weekly schedule

    c.  daily absenteeism

    d.  none of the above

19. Downtime is nonproductive time for equipment, which is expensive and unrecoverable. Downtime reported as mechanical or electrical downtime is useless. The best way to identify downtime is to report as three categories:

    a.  response to operations requests

    b.  perform work

    c.  awaiting materials

    What do each of these downtime categories signal?

    _____ quality of interdepartmental communications

    _____ adequacy of material control

    _____ effectiveness of repair actions

20. When is downtime attributable to operations?

    a.  equipment operators are absent or late

    b.  there are production delays

    c.  equipment is not returned to service promptly

    d.  all of the above

# CHAPTER 14

# Nonmaintenance Project Work

1. What does nonmaintenance project work include?

    a. construction

    b. equipment modification

    c. new equipment installation

    d. relocation of equipment to another location

    e. overhauls

2. What are the basic requirements needed to approve the conduct of nonmaintenance work?

    a. necessity

    b. feasibility

    c. proper funding

    d. production needs it

3. Why are ground rules necessary to control the amount of nonmaintenance work performed?

    a. excessive nonmaintenance work could interfere with the conduct of basic maintenance

    b. contractors may not be able to perform the work

    c. none of the above

4.  The following definition of maintenance specifically excludes certain nonmaintenance activities. What are they?

    Maintenance is the repair and upkeep of existing equipment, facilities, buildings, or areas in accordance with *current design specifications* to keep them in a safe, effective condition while meeting their intended purposes.

    a.  construction

    b.  modification

    c.  new equipment installation

    d.  relocating equipment

5.  Why must maintenance carefully monitor work done by contractors?

    a.  maintenance will maintain the equipment after the work is done

    b.  maintenance must understand how the work was done

    c.  maintenance should expect documentation on parts lists, diagrams, or instructions

    d.  all of the above

6.  Why should nonmaintenance work, once assigned to maintenance, be subjected to the same planning steps as maintenance?

    a.  the scope and nature of the work must be evaluated

    b.  the work must be prioritized to compete fairly with other major tasks

    c.  jobs should be considered on their merits

    d.  it precludes supervisors from accepting unapproved nonmaintenance work

7.  Approval of nonmaintenance projects must meet which of the following criteria?

    a.  be necessary and feasible

    b.  meet approval and funding criteria

    c.  be planned and properly engineered

    d.  have resources allocated by priority, when scheduled

    e.  all of the above

8.  Why are nonmaintenance projects considered a temporary activity that have a definite beginning and end?

    a.  they represent a temporary increase in the work load

    b.  they are weighed against future competing maintenance needs

    c.  all of the above

9.  Number the four project phases in correct order:

    _____ plan

    _____ define

    _____ complete

    _____ implement

10. By number, arrange the tasks of the project manager in logical order paralleling the phases of the project.

    _____ takes corrective actions as deviations occur

    _____ provides feedback to team members

    _____ resolves problems involving materials, supplies, or services

    _____ evaluates the project with an audit

    _____ monitors progress and measures against schedules and budgets

    _____ coordinates tasks of groups such as shops, maintenance, or engineering

    _____ disposes of surplus equipment materials

    _____ on project completion, writes documentation and manuals

    _____ trains personnel on use of the new equipment

    _____ reassigns project personnel

    _____ generates project report and management review

11. Anticipating the use of contractors suggests that the plant should consider whether the contractor will perform the following:

    a. only maintenance work

    b. maintenance plus capital nonmaintenance work

    c. plant services such as buildings and grounds work

    Match logical ways in which the contractor manages and executes the following work:

    _____ provides management, supervision, and planning

    _____ shows up at the proper times; plant administers and checks work

    _____ provides supervision and workers; plant provides management and planning

12. If a plant were to make a total transition from an in-house work force to the overall use of a contractor, number the phase-in steps required in logical order:

_____ engineering staff are provided by the contractor

_____ plant information system supports the effort

_____ contractor uses the plant warehouse

_____ contractor takes over the plant warehouse

_____ contractor provides only supervision and workers

_____ contractor replaces plant maintenance planning personnel

_____ contractor's information system links with the plant system

# CHAPTER 15

# Benchmarking

1. Benchmarking is a process of searching for best practices that lead to superior performance and then adapting them to improve one's own performance. This process can accrue several benefits including saving time, avoiding "trial and error," and speeding up the process of change. However, there is one essential caution, which is:

    a. benchmarked action must relate to actual improvement needs

    b. other plants may have a different program

    c. other plants may be organized differently

2. There are three major benefits derived from benchmarking: (1) cultural change, (2) performance improvement, and (3) human resources. Match the benefit to its purpose:

    _____ discovers gaps in performance and motivates organization to close gaps

    _____ motivates organization to find out why others do things differently or better

    _____ wants training to close gaps, bearing in mind what others have accomplished

3.  Benchmarking is a comparison of pertinent practices, not a contrast of performance. Why?

    a.  because the manner in which the target organization is doing things is the goal

    b.  because benchmarking dissimilar organizations can yield valuable ideas

    c.  because the target organization is a source of ideas, not a competitor

    d.  all of the above

4.  There are three types of benchmarking: (1) process, (2) performance, and (3) strategic. Match each with the most appropriate description:

    _____ a study of how the target organization implemented TPM throughout their plant

    _____ the manner in which Japan became a world leader in electronics

    _____ the measure of how effectively RCM was implemented

5.  Number the following benchmarking steps in logical order:

    _____ evaluate to determine improvement needs

    _____ develop an improvement plan, then prioritize needs

    _____ establish continuous improvement as a policy

    _____ implement new ideas and set goals for improvement

    _____ expand benchmarking to other industries or activities

    _____ measure improvement progress against goals

    _____ identify benchmarking as part of continuous improvement

    _____ "target" organizations that can help

    _____ conduct benchmarking with personnel who will implement ideas

    _____ incorporate new ideas developed into improvement plan

6. After a decision has been made to benchmark, the target organization should be carefully identified. What are some of the questions to be asked?

    a. do their activities fit your needs?

    b. do they have a similar maintenance organization?

    c. is their maintenance program similar to yours?

    d. are they in the same industry?

7. The key points listed in the left column should be considered when evaluating a target organization for benchmarking purposes. Match them with the reasoning listed in the right column.

    a. key personnel

    b. use of operators

    c. nonmaintenance

    d. value of information

    e. overall program

    f. adequate evaluation

    g. phony accomplishments

    h. meaningful differences

    _____ if they cannot document their program, look elsewhere

    _____ if there is no evaluation process, accomplishments may be speculation

    _____ modifications are expensed; yours are capitalized

    _____ if craftsmen don't report labor, costs may be inaccurate

    _____ your maintenance does construction and theirs does not; avoid comparisons

    _____ examine actions with less attention to job titles

    _____ why their operators do no maintenance but contribute to lower costs

    _____ confirm that their "claimed" accomplishments are worth adopting

8.  Benchmarking steps progress from simple to more complex. Number the steps listed in logical progression:

    _____ learn from successes

    _____ learn from mistakes

    _____ assess the "world-class" level

    _____ compare your plant with a leader in your industry

    _____ look beyond your industry

    _____ borrow any good ideas

    _____ apply internal best practices

# CHAPTER 16

# Material Control

1. Maintenance requires that the right materials in the proper quantity be available at the right time. Material control departments (purchasing and warehousing) provide procedures to identify stock and purchased materials in proper quantities and make them available to maintenance as requested. What are the specific actions that must be taken by either the maintenance (1) or material control (2) departments to ensure that these objectives are achieved? Insert number 1 or 2 for the following:

   _____ determine material needs

   _____ specify the quantity

   _____ determine time required

   _____ specify material control responsibilities for key maintenance personnel

   _____ obtain the materials required for planned work

   _____ obtain materials for all other types of work

   _____ provide sufficient lead time to procure materials

   _____ provide means of rebuilding major components

   _____ develop a forecast of major component replacements

2. What is the value of major component replacement forecasting?

    a. knowing advance need for materials allows material control to respond quickly

    b. generating standard bills of materials expedites ordering and procurement

    c. purchasing anticipates future needs and can ensure timely delivery

    d. warehouse can prepackage materials or tools for on-time delivery

    e. all of the above

3. Maintenance should avoid the role of the "professional" purchasing agent, but there are occasions when key maintenance personnel might have to assume an active role in materials procurement. What are these occasions?

    a. helping identify purchased materials

    b. making inquiries directly of suppliers to clarify parts identification

    c. having access to purchasing information to obtain the order status

    d. all of the above

4.  Often, local manufacturing of spare parts becomes necessary when an operation is in a remote location. What are some of the concerns you must be aware of to ensure that local manufacturing is carried out effectively?

    a.  field planners should coordinate with shop planners

    b.  warehouse or purchasing supervisors should work with shop planners

    c.  correct quantities should be determined to avoid costly machining setups

    d.  shops should not be used to manufacture parts when purchasing is cheaper

    e.  equipment standardization can reduce the need for manufactured parts

    f.  all of the above

5.  Number the steps involved in rebuilding major components in logical order:

    _____ warehouse places rebuilt components back in stock

    _____ transactions are controlled with a purchase order

    _____ maintenance tracks component performance in repair history

    _____ classified components are sent to local or vendor's shop

    _____ maintenance obtains parts from warehouse using work order

    _____ maintenance delivers components to warehouse for classification

    _____ vendors deliver the rebuilt components to the warehouse

6.  What are some of the advantages of using standard bills of materials to plan jobs that are repeated on a regular, periodic basis?

    a.  the planning task can be expedited

    b.  warehousing and purchasing can facilitate parts delivery

    c.  materials request can be forwarded electronically

    d.  warehousing and purchasing can better anticipate needs

    e.  all of the above

7. Material control success depends on a number of steps being carried out effectively.

   In the following list, insert "1" to identify actions taken by material control or "2" to identify actions requiring joint maintenance–material control.

   _____ ensure that material control procedures are effective and easy to follow

   _____ provide easy access to critical spare-parts lists

   _____ ensure that needed items and quantities are properly stocked

   _____ promptly replenish stock materials

   _____ keep the stockroom properly staffed

   _____ make stock withdrawal efficient

   _____ provide easy-to-use stock-return procedures

   _____ ensure proper stock accountability

   _____ make stock materials easy to identify

   _____ provide for interchangeability of parts

   _____ provide procedures for reserving stock parts

   _____ develop standard bills of materials for repetitive jobs

   _____ issue stock on presentation of approved work orders

   _____ summarize material costs by equipment and component

   _____ provide material costs job by job

   _____ provide for purchase-order tracking

   _____ provide drawings for remanufacture of parts

   _____ deliver materials to jobsites

8.  Control of crew tools is often delegated to warehouse personnel. What are some of the logical duties included in tool control?

    a.  develop accountability procedures to control issue and return

    b.  safeguard tools while in storage

    c.  ensure that tools are kept in good working condition

    d.  replace worn or damaged tools

    e.  recommend purchase of better tools

    f.  all of the above

9.  What are some considerations in making it easier for craftsmen or supervisors to obtain materials from the warehouse?

    a.  provide an easy way to identify needed materials

    b.  provide a means of graphic identification of materials

    c.  allow commonly used low-cost materials to be issued without paperwork

    d.  simplify work order procedure for material withdrawal

    e.  simplify procedures for returning unused materials

    f.  expedite procedures to shorten elapsed time for repairs

    g.  set up logical procedures for parts withdrawal during off-shifts

    h.  streamline procedures to obtain materials from warehouses

    i.  provide comprehensive, detailed instructions on all procedures

    j.  all of the above

10. The choice to replace production equipment at the end of its useful life cycle is an economic and performance decision requiring a team effort. Why?

    a. equipment cannot be selected only on a comparative cost basis

    b. equipment must be able to be operated effectively and easily maintained

    c. equipment must deliver required performance

    d. savings at purchase is otherwise canceled with lost production

    e. accounting should make the final decision

11. The following is a list of considerations involved in the decision to replace equipment. Mark each with either an "E" for economic considerations or a "P" for performance considerations.

_____ acquisition cost

_____ environmental considerations

_____ status of replacement parts

_____ technical support

_____ overhaul versus replacement

_____ utilization of equipment

_____ impact of inflation

_____ cash discounting

_____ financing

_____ tax considerations

_____ performance considerations

_____ depreciation

_____ cost of obsolescence

_____ ease of maintenance

_____ safety needs

_____ productive capacity

_____ operator training

_____ operating cost

_____ maintenance cost

_____ maintenance training

_____ new tools and procedures

# CHAPTER 17

# Cost Control

1.  Maintenance is the largest controllable cost in most plants and mines, often representing more than 35% of operating costs.

    □ True

    □ False

2.  After reading the following discussion on cost versus performance, answer the questions that follow.

    Total maintenance costs are controlled by the cost of material used rather than by the cost of labor, and most maintenance departments have a stable work force. Therefore, labor cost variations are usually the result of overtime. Material costs are determined by the volume of equipment that must be maintained and the speed at which equipment repair consumes spare parts and materials. Thus, equipment repairs characterized by repeated emergencies consume materials faster than equipment subjected to a deliberate, well-ordered maintenance program. Consequently, a key maintenance cost reduction objective should be to reduce emergency repairs (with better preventive maintenance) to slow the rate at which materials are consumed. Labor costs are incurred by the need to install materials. But emergency repairs are done less productively and more labor is consumed. The work is often rushed, resulting in inferior work quality. Emergency repairs are often done simply to get the equipment running again, and they must be repeated when equipment is needed less urgently. Thus, the expenditure of labor on emergencies is inefficient and wasteful because the work must often be redone.

The only direct control maintenance has over the cost of the work they perform is the efficiency with which they install materials. Therefore, better preventive maintenance leads to fewer emergencies and more planning results in better labor utilization and less labor cost to install materials. Maintenance can do little about a careless, poorly trained operator who damages equipment through improper operation in a remote part of the plant. As work is carried out, maintenance exercises better control over individual jobs that are planned and scheduled. These jobs, because they are better organized, are often completed in less elapsed downtime, consequently reducing the overall cost because less labor is used. Generally, material costs are the same whether the job is planned or not. Maintenance leaders make the greatest contribution to cost control through the productivity increases they bring about with quality work control.

    a.   What causes most labor cost variations?

    b.   Aside from equipment volume, what influences material cost?

    c.   Which type of work results in lower-quality work?

    d.   What is the most direct control maintenance has over cost?

    e.   Which type of work is most cost productive?

    f.   How can supervisors influence cost?

3.  At peak profitability, a certain plant showed a labor cost of $1.02 to install each dollar of material. If it now costs more than $4.50 to install the same amount of material, what are some of the underlying causes of this increase?

    a.  poor PM

    b.  inadequate planning

    c.  excessive emergency work

    d.  poor productivity

    e.  possible poor supervision

    f.  poor accounting practices

4.  The efficiency with which maintenance personnel install materials is a primary factor in their ability to control costs. What circumstances allow maintenance to increase the efficiency with which they install materials?

    a.  effective preventive maintenance finds problems before equipment fails

    b.  lead time provides more time to plan work

    c.  planned work is done with less manpower and less elapsed time

    d.  planned work is done more efficiently

    e.  efficiency results in higher-quality work

    f.  higher-quality work means that the repairs last longer

    g.  when repairs last longer, future repairs are done less frequency

    h.  less frequent repairs reduce the rate of material use

    i.  work efficiency reduces the amount and duration of labor use

    j.  hiring more people

5. Solid planning is necessary to enable personnel to install materials effectively and reduce or control costs. How is good planning possible?

    a. conduct a quality PM program

    b. preventive maintenance finds problems sooner

    c. allow adequate time to plan

    d. add more planners

    e. none of the above

6. Managers know that the cost of maintenance will not go down unless maintenance is done with fewer people and performed less often. What steps can managers take to safely reduce the cost of maintenance?

    a. improve productivity to enable work force reductions

    b. establish a better maintenance program with the goal of doing maintenance work less often

    c. use contractors and cut back on work force size

    d. order a work force reduction

7. Assuming a fixed work force level, reducing the cost of labor to install material is the best way to reduce overall maintenance costs.

    ☐ True

    ☐ False

8.  Do you agree with the following statement?

Effective maintenance cost reduction requires a strong detection-oriented PM program combined with good planning and scheduling. As PM inspections and testing uncover equipment problems sooner, problems will be simpler and less costly to correct. But these PM services will also find problems far enough in advance of failure to allow good planning to take place. Thus, every job that is planned and scheduled has the potential for being carried out more efficiently. That is, it will cost less in labor to install each dollar of material. The resulting improvement in productivity will then make labor available to do other work, possibly work now done by contractors. Herein are the savings. This fundamental relationship is essential to effective maintenance cost reduction.

□ Yes

□ No

9.  What are the characteristics of budgeting?

    a.  determination of the cost of 1 year's maintenance activity broken down into measurable elements

    b.  provides an opportunity to allocate resources to selected maintenance activities

    c.  a plan for controlling the rate and timing of allocated funds

    d.  a recommendation for approval of funds to support maintenance activities

    e.  a recommended plan of action for the expenditure of funds

    f.  a cost-reporting system provides a means of reporting expenditures with provision for the reallocation of funds or the allocation of additional funds for unexpected needs

10. The following paragraphs give definitions of three popular budgeting techniques: (1) zero-base budgeting, (2) performance budgeting, and (3) factored budgeting. Associate each of these techniques with the correct definition:

____ a. This budgeting technique relates the use of maintenance MH to production targets and then converts the relationship to labor and material costs to carry out the maintenance program.

____ b. The budgeting technique involves the allocation and control of resources through work activities. The activities are defined so that maintenance personnel can plan, schedule, control, and report on the work they do. Work activities must lend themselves to work–cost evaluation. Consequently, each significant activity should also relate to standardized resource requirements of labor, equipment use, material consumption, time requirement, and cost. Maintenance must be able to break down each significant activity into resource requirements and relate them to production rates and costs.

____ c. This budgeting technique begins by establishing activities that will be subjected to budgeting controls. Objectives are then set for these activities, with standards and performance methods established to meet the objectives. Next, priorities are set for funding the selected activities, and the activities compete with one another for funding. Thus, only the most important activities are funded and performance is measured on each. A continuing review procedure identifies programs that do not meet objectives so that funding can be cut back or eliminated.

11. Why would criteria to identify work requiring planning be helpful in maintenance cost reduction?

    a. some work does not require planning

    b. supervisors would know their work-control responsibilities

    c. planners can concentrate on planning more essential major jobs

    d. sending all jobs to the planners wastes time

    e. all of the above

12. Why should work be approved by line supervisors, not staff personnel?

    a. line supervisors have responsibility for work results

    b. staff personnel are often remote from actual field conditions

    c. staff personnel do not control maintenance crews

    d. all of the above

13. When scheduling planned jobs, why should production be included in decisions on timing of work, labor allocation, deferring work, or rejecting work if not considered necessary?

    a. the ultimate purpose of maintenance is production support

    b. the production plan should dictate when and how work is done

    c. production is responsible for utilization of maintenance

    d. all of the above

14. Senior managers are aware that maintenance costs are difficult to reduce unless maintenance is carried out with fewer people and less often. How can managers be best advised on actions that will avoid arbitrary cost-reduction actions while still preserving maintenance effectiveness?

    a. suggest concrete ways to improve productivity

    b. recommend different proven maintenance strategies

    c. review the existing program for reduction opportunities

    d. all of the above

15. Productivity must be improved to set the stage for using fewer personnel to conduct maintenance. What are some of the ways to improve productivity?

    a. insist that productivity be measured.

    b. ensure that the maintenance program is well defined and understood

    c. verify that the work load is correct

    d. carefully watch the utilization of labor

    e. perform evaluations to identify and prioritize improvement opportunities

    f. none of the above

16. What type of information provided to senior managers might prevent them from making arbitrary cost-reduction decisions?

    a. information on costs pinpointing necessary reductions

    b. performance indices that clearly identify problems

    c. factual cost-reduction recommendations

    d. all of the above

# CHAPTER 18

# Assessing Maintenance Performance

1. An evaluation is the first step of improvement to be taken in any maintenance improvement effort. Why?

    a. improvement needs must be clearly identified

    b. satisfactory performance areas should be left alone

    c. the right problems would be give priority

    d. improvement targets and their priorities can be established

    e. it will create the basis for a realistic improvement plan

    f. all of the above

2. What are the logical purposes of evaluations?

    a. to establish the current performance level

    b. to identify those activities needing improvement

    c. to spot activities being performed well

    d. to create a starting point for improvement efforts

    e. to provide opportunity to educate personnel

    f. to make improvement goals more attainable

    g. to generate pride in the organization

    h. all of the above

3.  What is the major reason that maintenance avoids evaluations?

    a.  maintenance dislikes nosy visitors

    b.  evaluations are misunderstood

    c.  performance is influenced by too many factors that maintenance can't control

4.  Of the evaluation techniques listed, which may be best for a maintenance organization that has traditionally shunned evaluations, fearing that the questionable performance of other departments will affect their outcome?

    a.  physical evaluation

    b.  questionnaire

    c.  physical evaluation with questionnaire

5.  What are some of the reasons that evaluations are more often welcomed than avoided?

    a.  people would rather do things right than wrong

    b.  troublesome activities are not identified and corrected

    c.  evaluations confirm need for help from others

    d.  evaluations may reveal incorrect use of maintenance resources

    e.  all of the above

6.  Few improvements can result unless there is an evaluation. Why?

    a.  causes of poor performance are not identified

    b.  good performance is seldom acknowledged

    c.  not possible to set improvement priorities

    d.  disruptive situations continue unchecked

    e.  poor, unworkable policies are perpetuated

    f.  all of the above

7. An evaluation announcement should:

    a. give dates

    b. state objectives

    c. avoid conflicts

    d. all of the above

8. To create a favorable environment for evaluations:

    a. explain why evaluations are helpful

    b. state positive results of evaluations

    c. identify what is to be evaluated

    d. schedule them regularly and continuously

    e. all of the above

9. Number the following randomly listed evaluation steps in logical order:

    _____ educate personnel on the purpose of the evaluation

    _____ schedule the maintenance evaluation

    _____ specify the dates of the next evaluation

    _____ publicize the content of the evaluation

    _____ use the most appropriate evaluation technique

    _____ take immediate action on the results

    _____ develop a policy for evaluations

    _____ provide advance notification of the evaluation

    _____ announce specific gains resulting from the evaluation

    _____ announce results

10. It is best to publicize evaluation content so maintenance can:

    a. prepare

    b. hide

    c. resist

    d. none of the above

11. What are the possible conflicts that could interfere in scheduling an evaluation?

    a. scheduled shutdowns

    b. peak vacation periods

    c. reorganization underway

    d. new systems being implemented

    e. maintenance manager indisposed

12. What situations constitute a favorable environment for maintenance?

    a. commitment from management to make maintenance successful

    b. cooperation from operations

    c. quality service levels from staff departments

    d. an effective work force

    e. good information system

13. Select the desirable characteristics of a self-evaluation questionnaire:

    a. participants represent whole operation

    b. factors influencing maintenance are evaluated

    c. personnel responding have personal knowledge

    d. results are quickly made available to all

    e. results are reviewed in detail

    f. others look at maintenance

    g. maintenance looks at others

    h. all of the above

14. What are some ways to convert evaluation results into improvements?

    a. use for education

    b. participation

    c. make recommendations

    d. test recommendations

    e. measure results

    f. all of the above

15. What are some of the parallel factors between equipment management and quality assurance that can make equipment management successful?

    a. good performance means higher-quality maintenance

    b. performance and quality are evident in equipment management

    c. the program emphasizes need for solid documentation

    d. documentation aids understanding and execution

    e. performance requires documentation to ensure quality

    f. all of the above

16. Following an unfavorable plant situation, a manager blamed maintenance for poor performance. The manager said, "We don't know what they don't know, and what we know may not be so." What did the manager mean?

    a. the manager was uncertain of the status of maintenance knowledge

    b. the manager was uncertain about their knowledge of maintenance

    c. the manager's attention to maintenance may have been neglected

    d. the manager recognized that maintenance might need more attention

    e. all of the above

17. Implementing quality into equipment management requires:

    a. determining quality objectives within the operating strategy

    b. policies for the efficient conduct of maintenance

    c. knowledge of the scope of production necessary to yield desired product quality

    d. proper documentation of the maintenance program

    e. identification of actions required to meet objectives

    f. all of the above

# Maintenance Performance Evaluation

The maintenance performance evaluation (MPE) is a self-evaluation technique that successfully identifies areas of maintenance that need improvement as well as those areas being performed satisfactorily. The results allow the plant or mine to develop an improvement program that identifies and prioritizes needed improvements. All plant or mine departments participate because successful maintenance requires the support and cooperation of all departments. Maintenance departments do not always have control over events and activities that affect their performance, such as the degree of cooperation from operations or the quality of material control by warehousing and purchasing. The MPE gives maintenance an opportunity to assess performance by other departments, and other departments have an opportunity to assess maintenance. Thus, any reluctance to evaluate is reduced or eliminated.

## UTILIZING THE MPE

The MPE is made up of 14 sections, each of which examines the critical elements that make maintenance management successful. Each section of the MPE is organized with an introduction followed by a group of standards. The introduction serves to explain the principles of the activity and set forth specific essential facts about the topic. The intent is to establish the framework within which individual participants will rate each standard.

To complete the evaluation, a cross section of not more than 25 personnel from each part of the operation would rate the performance standards. In a large operation—for example, a mine—the concentrator and refinery would each be evaluated separately. Each cross section is typically made up of 5 people from management and staff (manager, purchasing agent, warehouse manager, etc.), 8 from operations, and 12 from maintenance. Each group would consist of a vertical slice of the organization. The group of 12 from maintenance, for example, might include the maintenance manager, a general supervisor, a maintenance engineer, two maintenance planners, two maintenance supervisors, and five craftsmen from different skill groups. Smaller organizations could utilize fewer personnel.

Responses are structured so that personnel respond only to standards about which they have personal knowledge. Time to complete the evaluation is roughly 1½ hours. Standards are rated from 1 to 5 (highest). If a participant believes that the standard is important to his or her work but simply does not know how to rate it, an "X" is recorded for "I don't know." Generally, the ratings can be interpreted to mean:

1 — Never done or done poorly
2 — Rarely done or much improvement needed
3 — Occasionally done but trying to improve
4 — Done most of the time and continually improving
5 — Done all of the time, carried out professionally
X — It is important to me but "I don't know"
0 — Activity not applicable to me

A zero (0) rating means that the standard would not normally be encountered by a particular participant. To illustrate, an equipment operator could rate the standard as "preventive maintenance services are carried out with care and diligence by craft personnel"; however, the same equipment operator might not be able to rate whether "man-hours required, by craft, for the conduct of each preventive maintenance service have been established." Therefore, the operator would rate the standard zero (0) rather than from 1 to 5.

A judgment must be made by individuals whether they have sufficient personal knowledge to rate a standard objectively. In numerous applications of the MPE, it has been found that participants outside of maintenance were very knowledgeable about the level of maintenance performance and were discriminating and objective in rating only standards about which they had personal knowledge. Thus, no participant was excluded from considering any particular standard.

## THE MPE AS A PROGRESS MONITOR AND BENCHMARKING TOOL

The initial use of the MPE establishes the "as is" status of performance of a single operation such as a mine. MPE reports are used to develop an improvement plan by identifying activities in need of improvement. After the improvement effort is under way, repetitive use of the MPE can assess progress and the MPE can identify activities that have improved or require additional attention. The reader is encouraged to utilize the MPE as a basis for his or her own evaluation procedure and to apply it as deemed appropriate.

# Answers



1.  a–c. Reference: Page 1.

2.  a and c. Reference: Page 2.
    *Comment:* The company mission statement is assigned to all companies in
    the corporation and is used to guide plant managers in developing their
    production strategies for individual operations such as a mine. Terminology
    is better presented in the maintenance program definition.

3.  c. Reference: Page 3.

4.  *See the following.* Reference: Page 3.
    - 4 — plan work
    - 8 — measure work
    - 6 — assign work
    - 5 — schedule work
    - 7 — control work
    - 2 — identify work
    - 1 — request work
    - 9 — assess accomplishments
    - 3 — classify work

5.  Yes. Reference: Page 4.
    *Comment:* By assigning mutually supporting objectives in equipment
    management, it is the plant manager's intent to ensure cooperation among
    departments.

6.  a–d. Reference: Page 6.
    *Comment:* All of these potential problems are avoided by assigning com-
    plete, mutually supporting objectives to each department.

7.  a–d. Reference: Page 7.
    *Comment:* The policy demonstrates the plant manager's understanding of
    the value of preventive maintenance while providing guidance to mainte-
    nance on how they are to make it successful.

## CHAPTER 2 APPLYING THE PRINCIPLES OF EQUIPMENT MANAGEMENT

1. g. Reference: Page 11.

2. b and c. Reference: Pages 11–12.

3. c. Reference: Page 12.

4. a. Reference: Page 12.

5. *See the following.* Reference: Page 13.
   The maintenance program *defines* what maintenance must do, who will do *what*, how they will do it, and *why*.

6. a–d. Reference: Page 13.

7. a–c. Reference: Page 13.
   *Comment:* Measuring productivity is often avoided by maintenance because there are too many things affecting maintenance performance over which they have little control. For example, maintenance performance is hurt when purchasing does not deliver materials on time. Maintenance is dependent on quality purchasing service.

8. True. Reference: Page 13.
   *Comment:* The detection-oriented PM program successfully identifies significant equipment problems far enough in advance so that the resulting work can be planned. As a result, the work is carried out more efficiently.

9. a–c. Reference: Page 13.
   *Comment:* Some maintenance departments require planners to categorize all jobs before determining whether to plan them or have supervisors carry them out. When emergency repairs occur, any delay in this decision affects the supervisor's ability to react quickly to the emergency. A common solution is a jointly developed criteria understood and applied by supervisors who send only jobs requiring planning to planners.

10. a–d. Reference: Page 14.

11. True. Reference: Page 14.
    *Comment:* Joint scheduling ensures that the work will be done at a time that interferes least with operations and makes the best use of maintenance resources.

12. a–e. Reference: Page 14.

13. f. Reference: Page 14.

14. a, b. Reference: Page 14.

15.  a–d. Reference: Page 15.

16.  True. Reference: Page 15.
     *Comment:* Inclusion of all departments in the evaluation ensures the identi-
     fication and mutual correction of problems affecting joint performance.

## CHAPTER 3 DEVELOPING THE EQUIPMENT MANAGEMENT PROGRAM

1.   d. Reference: Pages 2–3.

2.   d. Reference: Page 17.

3.   *See the following.* Reference: Pages 17, 24–32.
     classifying — b
     requesting — a
     planning — b, c, d, and f
     measuring — b, e
     scheduling — b and a
     controlling — b, e, and f

4.   f. Reference: Page 18.

5.   e. Reference: Page 18.

6.   e. Reference: Page 19.

7.   e. Reference: Page 19.

8.   *Practical exercise.* Reference: Pages 26–27.

# CHAPTER 4 LEADERSHIP IN EQUIPMENT MANAGEMENT

1. a–e. Reference: Pages 34–35.

2. a–f. Reference: Page 36.

3. a–e. Reference: Pages 36–37.

4. a–c. Reference: Page 37.

5. a–c. Reference: Pages 37–38.

6. a–f. Reference: Pages 38–39.

7. a–e. Reference: Pages 39–40.

8. a–d. Reference: Page 40.

9. a–e. Reference: Pages 40–41.

## CHAPTER 5 ORGANIZATION

1. d. Reference: Page 43.

2. a–c. Reference: Page 43.

3. a–c. Reference: Pages 43–45.

4. d. Reference: Pages 43–45.

5. *See the following.* Reference: Pages 43–46.
   c — requests come to one point
   d — poor multicraft coordination
   a — one supervisor responsible
   b — performs specific tasks only

6. a–d. Reference: Pages 45–46.

7. a–d. Reference: Pages 46–47.

8. a, c, e, f, and g. Reference: Pages 47–49.

9. e. Reference: Page 50.

10. d. Reference: Pages 52.–53

11. a–d. Reference: Pages 52–53.

12. a–d. Reference: Pages 52–53.

13. b, c, and e. Reference: Pages 53–54.

14. a and b. Reference: Page 55.

# CHAPTER 6 WORK LOAD VERSUS WORK FORCE

1.  False. Reference: Page 57.
    *Comment:* The work load is the amount of essential work performed by maintenance.

2.  d. Reference: Page 57.

3.  *See the following.* Reference: Page 57.
    c — deliberately scheduled
    a — scheduled week by week
    b — immediate response
    d — fitted into existing work

4.  True. Reference: Page 57.

5.  *See the following.* Reference: Page 57.
    b — emergency repairs
    c — planned maintenance
    a — routine maintenance
    d — adjust work force when work load changes

6.  d. Reference: Page 58.

7.  *See the following.* Reference: Page 58.
    Maintenance is the *repair* and upkeep of *existing* equipment, facilities, buildings, or areas in accordance with *current* design specifications to keep them in a safe, effective *condition* while meeting their intended purposes.

8.  a, b, d, and e. Reference: Pages 58–59.

9.  b. Reference: Page 59.

10.  b. Reference: Page 59.

11.  b. Reference: Page 59.

12.  a–d. Reference: Page 60.

13.  a–f. Reference: Page 60.

14.  a, c, and d. Reference: Page 60.

15.  e. Reference: Page 60.

16.   e. Reference: Pages 60–61.

17.   c. Reference: Page 61.

18.   *See the following.* Reference: Page 66.
         a. 54 MH
         b. 319 MH
         c. 12%

## CHAPTER 7 IMPROVING WORK FORCE PRODUCTIVITY

1.  a. Reference: Page 69.

2.  b. Reference: Page 69.

3.  d. Reference: Page 69.

4.  d. Reference: Page 69.

5.  a. Reference: Page 69.

6.  f. Reference: Page 70.

7.  e. Reference: Page 71.

8.  d. Reference: Page 73.

9.  f. Reference: Page 73.

10.  e. Reference: Page 73.

11.  e. Reference: Page 73.

12.  g. Reference: Page 74.

13.  e. Reference: Page 74.

# CHAPTER 8 UNDERSTANDING PREVENTIVE MAINTENANCE

1. a and b. Reference: Page 79.
   *Comment:* When a unit of equipment requires an overhaul, there are so many things that are faulty or inoperable that the unit must be removed from service and all components removed and replaced with new or rebuilt components. There is nothing left to prevent. Therefore, overhauls are not preventive maintenance but rather planned and scheduled major maintenance.

2. d. Reference: Page 79.

3. b. Reference: Page 79.

4. *See the following.* Reference: Page 79.
   A fixed interval is — a.
   A variable interval is — d.

5. d. Reference: Page 79.

6. b. Reference: Page 79.

7. f. Reference: Page 80.

8. *See the following.* Reference: Pages 80–81.
   shock-pulse diagnosis — C
   cleaning — R
   testing and calibration — R
   ultrasonic testing — C
   inspection — R
   infrared scanning — C

9. d. Reference: Page 82.

10. *See the following.* Reference: Page 84.
    emergency repairs: decrease
    planned maintenance: increase
    unscheduled repairs: decrease

11. *See the following.* Reference: Page 85.
    organizing steps: b, d, f
    operating steps: a, c, e

12. *See the following.* Reference: Page 86.
    planned work: a, b,
    unscheduled repairs: d
    emergency repairs: c

13. *See the following.* Reference: Pages 88–89.

> 4 — planned work is completed in less elapsed downtime
> 3 — emergency work is avoided by early discovery of problems
> 7 — equipment is put back into production sooner
> 6 — planned work is done more deliberately with greater quality
> 9 — profitability is improved
> 5 — planned work is completed with less labor
> 1 — PM services find equipment problems in time to plan work
> 2 — resulting deficiencies can be converted into planned work
> 8 — planned work is completed at less overall cost

14. f. Reference: Page 89.

15. *See the following.* Reference: Pages 91–97.

    a. vibration monitoring, lube and fuel analysis, wear particle analysis, bearing temperature analysis, performance monitoring
    b. insulation resistance, motor current signature, motor circuit, polarization index
    c. infrared thermography, visual inspection

## CHAPTER 9 EFFECTIVE PLANNING AND SCHEDULING

1. *See the following.* Reference: Page 99.
   a. pre-organize selected major jobs — P
   b. identify resources needed — P
   c. ensure that work can be done in the least amount of downtime — P
   d. ensure the most effective use of resources — P
   e. complete at the lowest possible cost — P
   f. determine best time to complete the work — S

2. c. Reference: Page 99.

3. c. Reference: Page 99.

4. f. Reference: Page 100.

5. e. Reference: Page 101.

6. *See the following.* Reference: Pages 102–104.
   15 — conduct the scheduling meeting
   4 — determine if standards apply
   11 — open work order and order materials and shop work
   5 — confirm the job scope
   6 — make the job plan and set up the work order
   13 — confer with operations on job timing
   1 — identify the work
   7 — determine resources
   8 — establish labor by craft
   2 — investigate
   9 — estimate cost, set the job priority, and get approval
   10 — estimate preliminary time to do job
   12 — await receipt of materials
   3 — get advice from the crew
   14 — arrange for rigging, transport, and tools
   16 — monitor job execution and note cost and performance

7. h. Reference: Pages 104–105.

8. e. Reference: Pages 104–105.

9. b. Reference: Page 105.

10. f. Reference: Page 105.

11. a–c. Reference: Page 105.

12. j. Reference: Pages 105–106.

13. e. Reference: Pages 106–107.

14. *See the following.* Reference: Pages 108–111.
    3 — distribute the approved schedule
    2 — allocate labor
    4 — check performance indices
    2 — confirm priorities
    3 — explain the tasks
    1 — prepare preliminary plan
    3 — confirm shutdown times
    1 — determine completeness of planning
    3 — confirm use of equipment
    1 — discuss within maintenance
    2 — negotiate with operations
    3 — confirm delivery of materials
    1 — verify availability of materials
    5 — visit field location to observe work in progress or problems
        developing
    2 — obtain schedule approval
    1 — verify completion of shop work
    4 — check schedule compliance
    1 — propose shutdown times
    1 — verify work force

15. a and b. Reference: Page 112.
    *Comment:* Logically, PM services requiring equipment shutdown should
    be included in the weekly schedule. However, PM services that can be
    performed while the equipment is running are best scheduled by the
    maintenance supervisor who has a better picture of the availability of the
    equipment and the labor resources available.

16. d. Reference: Page 113.
    *Comment:* After the accumulation of a certain number of operating hours
    or the passage of time, the component may still have useful life. Inspect
    the component to determine its actual condition before replacing it, and
    verify, if possible, the cause of failure. Also examine recent repair history to
    identify events leading up to failure.

17. a. Reference: Page 113.

18. e. Reference: Pages 113–117.

## CHAPTER 10 RELIABILITY CENTERED MAINTENANCE

1. f. Reference: Page 119.

2. f. Reference: Page 119.

3. c. Reference: Page 119.

4. d. Reference: Page 119.

5. b. Reference: Page 120.

6. e. Reference: Page 121.

7. *See the following.* Reference: Page 121.
   a. Doesn't perform functions to performance standard specified — F
   b. An indication that the failure process has started — P

8. d. Reference: Page 122.

9. a. Reference: Page 122.

10. a. Reference: Page 122.

11. e. Reference: Page 122.

12. *See the following.* Reference: Pages 123–126.
    2 — identify the functions of the equipment
    4 — determine the types of failures
    5 — enumerate the consequences of failures
    1 — select the most critical equipment
    6 — rank the consequences of failures
    8 — incorporate into the overall maintenance plan
    3 — establish performance standards
    7 — apply the most effective condition-monitoring techniques

13. g. Reference: Page 126.

14. b. Reference: Page 126.

15. a and b. Reference: Page 127.

16. True. Reference: Page 127.

17. a. Reference: Page 127.

18.  *See the following.* Reference: Pages 128–129.
      failure to detect problem — b
      operating error — d
      material fatigue — a
      infant mortality — c
      random failures — e

19.  e. Reference: Page 130.

20.  h. Reference: Pages 130–131.

21.  e. Reference: Page 131.

22.  True. Reference: Page 131.

23.  h. Reference: Page 131.

24.  True. Reference: Page 133.

25.  a–c. Reference: Page 133.

26.  a–d. Reference: Page 133.

27.  *See the following.* Reference: Pages 133–136.
      RCFA — c
      FMEA — a
      RBI — b

28.  *See the following.* Reference: Page 136.
      T — total cost of reliability is the cost of failures not averted
      T — failures result in injuries, lost product, and equipment damage
      T — maintenance programs act to reduce the cost of maintenance
      T — successful RCM attacks the cost of failure
      T — failure avoidance increases reliability

## CHAPTER 11 TOTAL PRODUCTIVE MAINTENANCE

1. f. Reference: Page 139.

2. a–d. Reference: Page 140.

3. a–c. Reference: Page 142.

4. d. Reference: Page 142.

5. f. Reference: Page 142.

6. a. Reference: Page 143.

7. a–f. Reference: Page 143.

8. a and b. Reference: Pages 143–144.
   *Comment:* RCM emphasizes the use of condition monitoring to avoid random failures.

9. a–e. Reference: Page 145.

10. Paint A. Reference: Pages 145–146.
    *Comment:* Using paint A would cost $50,000 for 6 years versus $35,000 for paint B during the same period.

11. *See the following.* Reference: Pages 146–147.
    2 — initiate autonomous maintenance
    3 — increase skills of operators and maintainers
    1 — reduce the six big losses
    4 — apply these factors to specific equipment

12. *See the following.* Reference: Page 147.
    preparatory — b
    implementation — c
    stabilization — a

# CHAPTER 12 IMPLEMENTING INFORMATION SYSTEMS

1.  e. Reference: Page 151.

2.  h. Reference: Page 151.

3.  c. Reference: Pages 151–152.

4.  c. Reference: Page 152.

5.  f. Reference: Page 152.

6.  e. Reference: Pages 152–153.

7.  b. Reference: Page 153.

8.  *See the following.* Reference: Pages 156–160.
    - 4 — load all files and confirm equipment numbering
    - 5 — verify that hardware and networking arrangements are functional
    - 8 — conduct effective, realistic training
    - 6 — verify field data source accuracy
    - 2 — ensure that roles of key personnel are correct and understood
    - 3 — phase out all previous, conflicting procedures
    - 7 — establish a core training group
    - 1 — confirm the soundness of the equipment management program
    - 9 — develop implementation schedule with specific objectives
    - 10 — monitor system use and accomplishments

9.  False. Reference: Page 160.

10.  f. Reference: Pages 160–161.

11.  l. Reference: Page 161.

12.  *See the following.* Reference: Pages 161–162.
    - maintenance work order (MWO) — a
    - maintenance work request (MWR) — b
    - verbal orders — c
    - standing work order (SWO) — d
    - engineering work order (EWO) — e

13.   i. Reference: Page 164.

14.   *See the following.* Reference: Page 164.
>     A — time card
>     W — SWO
>     P — tons of throughput
>     A — stock issue card
>     W — EWO
>     A — purchase order
>     W — MWO
>     P — operating hours

15.   e. Reference: Pages 164–165.

16.   d. Reference: Page 165.

17.   d. Reference: Page 166.

18.   b and d. Reference: Page 167.

19.   d. Reference: Page 168.

# CHAPTER 13 ESSENTIAL INFORMATION

1. c and d. Reference: Page 171.

2. *See the following.* Reference: Pages 171–172.
   maintenance engineers — e
   general supervisor — c
   planner — b
   supervisor — a
   craftsmen — d

3. a–c. Reference: Page 172.

4. a–c. Reference: None.
   *Comment*: Cost and repair history are among the information elements that signal the work that mush be done. Reduced costs and fewer repair history entries confirm that actions taken were necessary. Less downtime suggests that equipment is more reliable as a result of greater work effectiveness.

5. a–c. Reference: Page 172.

6. *See the following.* Reference: Page 173.
   a. 244 MH
   b. 233 MH
   c. 277 MH
   d. 16%
   e. 1,967 MH

7. a–c. Reference: Page 175.

8. *See the following.* Reference: Page 176.
   a. +60 MHE (MHE = man-hours estimated)
   b. Riggers (RIGR)
   c. +51 MHE (ELEC)
   d. Fitters (FITR)
   e. 723 MHE
   f. +113 MHE

9. *See the following.* Reference: Page 177.

**Backlog Data**

| | | | | | | | | | | |
|---|---|---|---|---|---|---|---|---|---|---|
| Mechanic | 1,270 | 1,490 | 1,840 | 1,950 | 2,025 | 1,920 | 2,000 | 1,910 | 1,495 | 1,340 |
| Millwright | 1,455 | 1,545 | 1,660 | 1,705 | 1,780 | 1,750 | 1,775 | 1,730 | 1,530 | 1,455 |
| Electrician | 710 | 750 | 790 | 805 | 800 | 800 | 805 | 790 | 740 | 710 |
| Total | 3,435 | 3,785 | 4,290 | 4,460 | 4,605 | 4,470 | 4,580 | 4,430 | 3,765 | 3,505 |

10. *See the following.* Reference: Page 178.
    a. craft 2
    b. craft 3

11. *See the following.* Reference: Pages 178–180.
    a. $3,495
    b. $2,100
    c. 125 MH
    d. Week 19, 2007
    e. 34338A
    f. 90

12. *See the following.* Reference: Page 183.
    a. Component 01, engine
    b. $173,997
    c. 1920 MH of regular time and overtime
    d. $15,457
    e. A "general" component encourages data dumping

13. *See the following.* Reference: Page 183–184.
    a. 183 operating hours (OpHrs)
    b. 12 weeks
    c. 6.6 MH
    d. MWO #65664
    e. Failure code (FC) #2

14. c. Reference: Page 187.

15. *See the following.* Reference: Page 187.
    2 — assurance that funds are well-spent and quality work is done
    1 — trends and summaries to assess maintenance effectiveness
    3 — identify work, control work, and measure accomplishments

16. *See the following.* Reference: Page 187.
    b — how well maintenance contributes to reduction of downtime
    a — overall maintenance contribution to plant productive effort
    c — reveals how effectively maintenance uses its labor to perform work

17. e. Reference: Page 187.

18. a and b. Reference: Page 188.

19. *See the following.* Reference: Page 189.
    a — quality of interdepartmental communications
    c — adequacy of material control
    b — effectiveness of repair actions

20. d. Reference: Page 189.

# CHAPTER 14 NONMAINTENANCE PROJECT WORK

1.  a–d. Reference: Page 199.

2.  a–c. Reference: Page 199.

3.  a. Reference: Pages 200–201.

4.  a and b. Reference: Page 201.
    *Comment: Existing* suggests that maintenance cannot begin until the new equipment, buildings, and facilities are created through construction activities. *Current design specifications* suggests that modifications are not maintenance.

5.  d. reference: Page 201.

6.  a–d. Reference: Page 201.

7.  e. Reference: Page 202.

8.  c. Reference: Page 204.

9.  *See the following.* Reference: Pages 204–205.
    2 — plan
    1 — define
    4 — complete
    3 — implement

10. *See the following.* Reference: Page 205.
    3 — takes corrective actions as deviations occur
    4 — provides feedback to team members
    5 — resolve problems on materials, supplies, or services
    10 — evaluates project with audit
    2 — monitors progress and measures against schedules and budgets
    1 — coordinates tasks of groups like shops, maintenance, or engineering
    9 — disposes of surplus equipment materials
    6 — on projection completion, writes documentation and manuals
    7 — trains personnel on use of the new equipment
    8 — reassigns project personnel
    11 — generates project report and management review

11. *See the following.* Reference: Pages 206–207.
    b — provides management, supervision, and planning
    c — shows up at the proper times; plant administers and checks work
    a — provides supervision and workers; plant provides management and planning

12.  *See the following.* Reference: Page 208.

> 3 — engineering staff are provided by the contractor
> 4 — plant information system supports the effort
> 6 — contractor uses the plant warehouse
> 7 — contractor takes over the plant warehouse
> 1 — contractor provides only supervision and workers
> 2 — contractor replaces plant maintenance planning personnel
> 5 — contractor's information system links with the plant system

# CHAPTER 15 BENCHMARKING

1.  a. Reference: Page 211.

2.  *See the following.* Reference: Page 211.
    2 — discovers gaps in performance and motivates the organization to close gaps
    1 — motivates organization to find out why others do things differently or better
    3 — wants training to close gaps, bearing in mind what others have accomplished

3.  b. Reference: Page 212.

4.  *See the following.* Reference: Pages 212–213.
    2 — a study of how the target organization implemented TPM throughout their plant
    3 — the manner in which Japan became a world leader in electronics
    1 — the measure of how effectively RCM was implemented

5.  *See the following.* Reference: Pages 213–214.
    3 — evaluate to determine improvement needs
    4 — develop an improvement plan, then prioritize needs
    1 — establish continuous improvement as a policy
    8 — implement new ideas and set goals for improvement
    10 — expand benchmarking to other industries or activities
    9 — measure improvement progress against goals
    2 — identify benchmarking as part of continuous improvement
    5 — identify "target" organizations that can help
    6 — conduct benchmarking with personnel who will implement ideas
    7 — incorporate new ideas developed into improvement plan

6.  a–c. Reference: Page 215.

7.  *See the following.* Reference: Pages 215–216.
    e — if they cannot document their program, look elsewhere
    f — if there is no evaluation process, accomplishments may be speculation
    c — modifications are expensed; yours are capitalized
    d — if craftsmen don't report labor, costs may be inaccurate
    h — your maintenance does construction and theirs does not; avoid comparisons
    a — examine actions with less attention to job titles
    b — why their operators do no maintenance but contribute to lower costs
    g — confirm that their "claimed" accomplishments are worth adopting

8.  *See the following.* Reference: Page 218.

>    2 — learn from successes
>    1 — learn from mistakes
>    7 — assess the "world-class" level
>    5 — compare your plant with a leader in your industry
>    6 — look beyond your industry
>    3 — borrow any good ideas
>    4 — apply internal best practices

# CHAPTER 16 MATERIAL CONTROL

1. *See the following.* Reference: Page 219.

    1 — determine material needs
    1 — specify the quantity
    1 — determine time required
    2 — specify material control responsibilities for key maintenance personnel
    1 — obtain the materials required for planned work
    1 — obtain materials for all other types of work
    1 — provide sufficient lead time to procure materials
    2 — provide means of rebuilding major components
    1 — develop a forecast of major component replacements

2. e. Reference: Page 219.

3. d. Reference: Page 221.

4. f. Reference: Page 221.

5. *See the following.* Reference: Page 222.

    5 — warehouse places rebuilt components back in stock
    3 — transactions are controlled with a purchase order
    7 — maintenance tracks component performance in repair history
    2 — classified components are sent to local or vendor's shop
    6 — maintenance obtains parts from warehouse using work order
    1 — maintenance delivers components to warehouse for classification
    4 — vendors deliver the rebuilt components to the warehouse

6. e. Reference: Page 222.

7. *See the following.* Reference: Pages 222–223.

    1 — ensure that material control procedures are effective and easy to follow
    1 — provide easy access to critical spare-parts lists
    1 — ensure that needed items and quantities are properly stocked
    1 — promptly replenish stock materials
    1 — keep the stockroom properly staffed
    1 — make stock withdrawal efficient
    1 — provide easy-to-use stock-return procedures
    2 — ensure proper stock accountability
    2 — make stock materials easy to identify
    2 — provide for interchangeability of parts
    2 — provide procedures for reserving stock parts
    2 — develop standard bills of materials for repetitive jobs
    2 — make issues on presentation of approved work orders
    2 — summarize material costs by equipment and component

    2 — provide material costs job by job
    2 — provide for purchase-order tracking
    2 — provide drawings for remanufacture of parts
    2 — deliver materials to jobsites

8.  f. Reference: None (general knowledge and appreciation).

9.  j. Reference: Page 223.

10.  a–d. Reference: Page 224.

11.  *See the following.* Reference: Pages 225–233.
    E — acquisition cost
    P — environmental considerations
    P — status of replacement parts
    P — technical support
    P — overhaul versus replacement
    P — utilization of equipment
    E — impact of inflation
    E — cash discounting
    E — financing
    E — tax considerations
    P — performance considerations
    E — depreciation
    E — cost of obsolescence
    P — ease of maintenance
    P — safety needs
    P — productive capacity
    P — operator training
    P — operating cost
    P — maintenance cost
    P — maintenance training
    P — new tools and procedures

## CHAPTER 17 COST CONTROL

1. True. Reference: Page 235.

2. *See the following.* Reference: Page 235.
   a. Overtime
   b. Speed at which equipment repair consumes materials
   c. Emergency repairs
   d. The efficiency with which they install materials
   e. Planned and scheduled work
   f. They can bring about productivity increases through quality work control

3. a–e. Reference: Page 236.

4. a–i. Reference: Page 236.

5. a–c. Reference: Page 236.

6. a and b. Reference: Page 236.

7. True. Reference: Page 236.

8. Yes. Reference: Page 236.

9. a–e. Reference: Page 236.
   *Comment:* The cost-reporting system is a part of the regular information system, but it supports fiscal management of maintenance budgeting along with many other functions.

10. *See the following.* Reference: Pages 236–244.
    3 — a
    2 — b
    1 — c

11. e. Reference: Page 243.

12. d. reference: Page 243.

13. d. Reference: Page 243.

14. d. Reference: Pages 243–246.

15. a–e. Reference: Pages 243–246.

16. d. Reference: Page 244.

## CHAPTER 18 ASSESSING MAINTENANCE PERFORMANCE

1.  f. Reference: Page 249.

2.  h. Reference: Page 249.

3.  c. Reference: Page 249.

4.  c. Reference: Page 251.

5.  e. Reference: Page 251.

6.  f. Reference: Page 251.

7.  d. Reference: Page 251.

8.  e. Reference: Page 251.

9.  *See the following.* Reference: Pages 251–258.
    - 3 — educate personnel on the purpose of the evaluation
    - 4 — schedule the maintenance evaluation
    - 10 — specify the dates of the next evaluation
    - 5 — publicize the content of the evaluation
    - 6 — use the most appropriate evaluation technique
    - 8 — take immediate action on the results
    - 1 — develop a policy for evaluations
    - 2 — provide advance notification of the evaluation
    - 9 — announce specific gains resulting from the evaluation
    - 7 — announce results

10.  a. Reference: Page 253.

11.  a–c. Reference: Page 253.

12.  a–c. Reference: Page 254.

13.  h. Reference: Page 255.

14.  f. Reference: Page 258.

15.  f. Reference: Page 259.

16.  e. Reference: Page 258.

17.  f. Reference: Page 264.